CliffsNotes® STAAR® EOC Algebra I Quick Review

By Jorge A. Jacquez

About the Author

Jorge A. Jacquez is an acclaimed educator and curriculum writer in the state of Texas. Each year, Mr. Jacquez lectures at local and nationally recognized conferences on best teacher practices and modern techniques for nurturing algebraic reasoning. He began his professional teaching career in his hometown of El Paso, Texas, and is currently an Algebra I, Geometry, and Precalculus teacher in Austin, Texas.

Editorial

Executive Editor: Greg Tubach

Senior Editor: Christina Stambaugh

Copy Editor: Lynn Northrup

Technical Editors: Mary Jane Sterling and Tom Page

Proofreader: Donna Wright

CliffsNotes® STAAR® EOC Algebra I Quick Review

Library of Congress Control Number: 2015943211
ISBN: 978-1-957671-12-3

Printed in the United States of America
DOC 10 9 8 7 6 5 4 3 2 1

Table of Contents

INTRODUCTION

CliffsNotes STAAR EOC Algebra I Quick Review is a reference tool that will help you review the important elements of Algebra I necessary to master the STAAR End of Course (EOC) Exam.

What Is STAAR End of Course?

STAAR stands for "State of Texas Assessments of Academic Readiness," and *EOC* stands for "end of course." Texas high school students must meet a Satisfactory Academic Performance (Level II) on each EOC assessment they take in order to graduate.

Students will be tested in five reporting categories. These are listed below, along with the corresponding chapters in this book:

- Number and Algebraic Methods (chapters 1, 7, and 8)
- Describing and Graphing Linear Functions, Equations, and Inequalities (chapters 2, 4, and 5)
- Writing and Solving Linear Functions, Equations, and Inequalities (chapters 2, 3, 4, and 5)
- Quadratic Functions and Equations (chapters 6 and 9)
- Exponential Functions and Equations (chapters 7 and 8)

Important Facts About the STAAR EOC Algebra I Exam

- The exam consists of 54 multiple-choice questions.
- Students will have 4 hours to complete the STAAR EOC Algebra I Exam. Students are allowed to take breaks to get a drink of water, have a snack, or use the restroom, but the test clock will not pause for these breaks. Practice timing yourself while taking the two

full-length practice tests in chapters 10 and 11 to prepare for the timed setting of the actual exam.

- Students will have access to a handheld, four-function, scientific, or graphing calculator. There will be at least one calculator for every five students on the day of the test. If a calculator is used by multiple students, its memory must be cleared after each use.
- Your STAAR EOC Algebra I test booklet contains reference materials, including a formula chart and graph paper. On pages 221–222, we've provided a formula chart similar to the one that will be in your test booklet. Familiarize yourself with these formulas.
- The test may be administered as a paper-and-pencil exam or on a computer. The test administration method varies by school and by school district. Some school districts administer this exam toward the end of the school year. The results of this exam are sometimes used to award credit for an Algebra I course.

For additional information on the STAAR EOC Algebra I Exam, visit http://tea.texas.gov/student.assessment/staar/math/.

How to Use This Book

Suggested approaches

There are a number of ways you can use this book to prepare for the STAAR EOC Algebra I Exam. You decide what works best for your needs. Here are a few suggested approaches:

- **Approach 1: (beginner)**
  - Read chapters 1–9.
  - Answer the Chapter Check-Out questions at the end of each chapter to make sure you understand the content.
  - Take Practice Test 1 (chapter 10) for practice.
  - Take Practice Test 2 (chapter 11) for additional practice.
- **Approach 2: (intermediate)**
  - Take Practice Test 1 (chapter 10) as a diagnostic test to see what areas you need to focus on most. Check your answers and identify the topics of the questions you missed.
  - Review these topics in chapters 1–9.

    - Answer the Chapter Check-Out questions at the end of each chapter to make sure you understand the content.
    - Take Practice Test 2 (chapter 11) for final practice.
- **Approach 3: (advanced)**
    - Answer the Chapter Check-Out questions at the end of each chapter. Review the related chapter content for any questions you missed.
    - Take Practice Test 1 (chapter 10) for practice.
    - Take Practice Test 2 (chapter 11) for additional practice.

Tech tips

In several chapters in this book, we've provided Tech Tips to show you how to use a calculator to solve certain types of problems. Practice using your calculator where appropriate to check your solutions and as a time-saver.

Chapter 1

FOUNDATIONS FOR ALGEBRA

Chapter Check-In

- ❑ Working with numbers
- ❑ Variables and expressions
- ❑ Evaluating expressions
- ❑ Order of operations
- ❑ Combining like terms
- ❑ Independent and dependent variables
- ❑ Properties of numbers
- ❑ Real number system
- ❑ Coordinate plane
- ❑ TECH TIPS: Evaluating and simplifying expressions using a graphing calculator

Before we dive into the basic elements of algebra, let's first discuss some of its essential building blocks. In order to use algebra to solve real-world problems, you must be able to understand its fundamental rules. We will also explore the language of algebra, which will allow us to switch back and forth between algebra and words.

Working with Numbers

A **variable** is an unknown numerical value, usually represented by a symbol or a letter of the alphabet.

A **constant** is a number that does not change and has no variable.

A **mathematical operation** is an action or a process that will sometimes result in a new value. The four main mathematical operations are addition, subtraction, multiplication, and division. These mathematical operations can also be represented by symbols. They are:

Addition	Subtraction	Multiplication	Division
$+$	$-$	$\times$	$\div$

There are other mathematical operations, such as squaring a number or finding its square root.

Numerical expressions can be written by combining constants and mathematical operation symbols, as shown below.

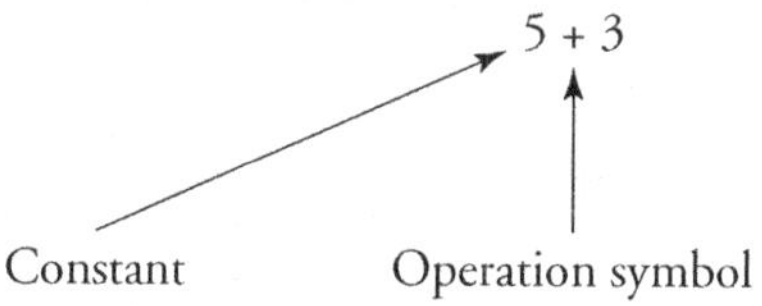

Variables and Expressions

Algebraic expressions contain constants, mathematical operation symbols, and variables. Here is an example of an algebraic expression:

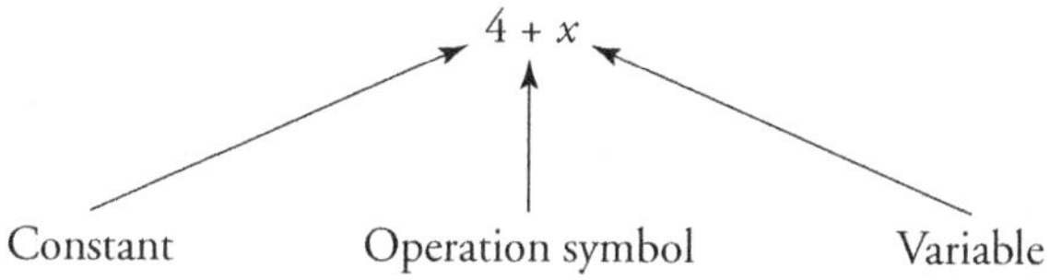

This expression is read as *four plus x*. It can also be read as *the sum of four and x*, or *x more than four*.

Often, keywords indicate the mathematical operation that is taking place. Here is a list of some of the most common keywords and their corresponding mathematical operations.

+	–	×	÷
plus	minus	times	divided by
add	subtract	multiply	divided into
more	less	factor of	out of
sum	difference	product	quotient
increase	decrease	twice	ratio
combined	fewer than	tripled, etc.	half, third, etc.

There is more than one way to show certain mathematical operations, such as multiplication and division.

For example, the expression *five times seven* can be represented as follows:

5×7 Multiplication sign

$5 \cdot 7$ Dot product

$5(7)$ Implied multiplication, parentheses

$5[7]$ Implied multiplication, brackets

The expression *3 times p* can also be written using implied multiplication without parentheses as $3p$; p represents an unknown value, and the number 3 that is multiplied by the variable is known as the **coefficient.**

The expression x *divided by* z can be represented as follows:

$x \div z$	$z\overline{)x}$	$\dfrac{x}{z}$
Obelus (division symbol)	Long division symbol	Fraction/ratio

Example 1:

Write a mathematical expression for the statement *nine less than a number x.*

Solution:

$x - 9$

Note that the expression $9 - x$ would be incorrect and would actually be read as *a number x less than 9.*

Example 2:

Write a mathematical expression for the statement *the difference of ten and the sum of three and a number p.*

Solution:

$10 - (3 + p)$

Example 3:

Write the following algebraic expression in words two different ways: $\frac{q}{8}$.

Solution:

Possible solutions include the following:

the quotient of q and 8
the ratio of q to 8
q divided by 8 (Note that the numerator, q, comes first in the phrase and the denominator, 8, comes second.)

Evaluating Expressions

To **evaluate** an expression means to calculate its numerical value. To evaluate an algebraic expression, which includes variables, you need to **substitute** numbers in place of the variables before simplifying the expression.

Example 4:

Evaluate the expression $20 \div 5$.

Solution:

4, divide 20 by 5

Example 5:

Evaluate each expression for $a = 2$, $b = 8$, $c = -5$.

A. $a \cdot c$

B. $b - c$

C. $\frac{b}{a}$

Solution:

A. $a \cdot c = (2)(-5) = -10$

B. $b - c = 8 - (-5) = 13$

C. $\frac{b}{a} = \frac{8}{2} = 4$

Order of Operations

When evaluating an expression, mathematical operations should always be performed in a particular order to ensure that solutions are always consistent. We refer to this as the **order of operations.**

Acronyms are commonly used to recall the order of operations. One such acronym is **GEMDAS,** which stands for

G	Grouping symbols
E	Exponents
M **D**	Multiplication OR Division (in order from left to right)
A **S**	Addition OR Subtraction (in order from left to right)

You might also see the acronym "PEMDAS," where "P" stands for "parentheses (and other grouping symbols)." These terms are interchangeable.

There are multiple types of **grouping symbols;** the most common are listed below:

( )	Parentheses
[ ]	Brackets
$\sqrt{\ }$	Radical sign
$\frac{a}{b}$	Fraction bar (the numerator and denominator are separate groups)
| |	Absolute value symbol

Example 6:

Evaluate $11 - 8 \div 4$.

Solution:

$$11-8\div4$$
$$11-(2) \quad \text{Division}$$
$$9 \quad \text{Subtraction}$$

In algebra, you will often have to **substitute** a number for a variable and simplify using the order of operations to evaluate an expression.

Example 7:

Evaluate the expression $10 - x^2 + 3(5)$ for $x = 2$.

Solution:

$$10-(2)^2+3(5)$$
$$10-(4)+3(5) \quad \text{Exponents}$$
$$10-4+15 \quad \text{Multiplication}$$
$$6+15 \quad \text{Addition/Subtraction (from left to right)}$$
$$21 \quad \text{Addition}$$

Combining Like Terms

All terms with the same variable raised to the same power are called **like terms.**

Consider a trip to the grocery store in which you buy 3 oranges and your friend buys 2 oranges. Together, you can say you have a total of 5 oranges.

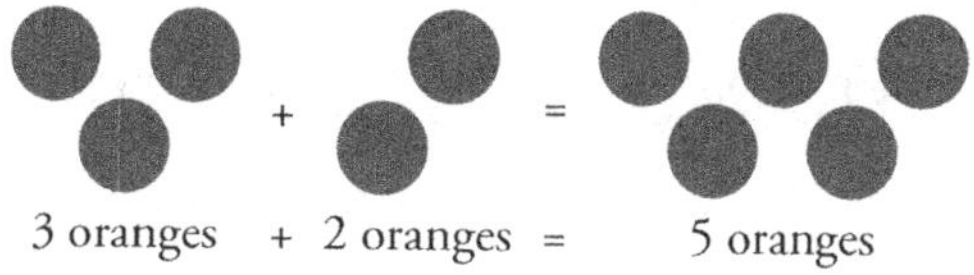

However, if you purchase 3 oranges and your friend decides to buy 2 limes, you cannot combine those numerical values because the two items are unlike. Therefore, you would say that together you have 3 oranges and 2 limes.

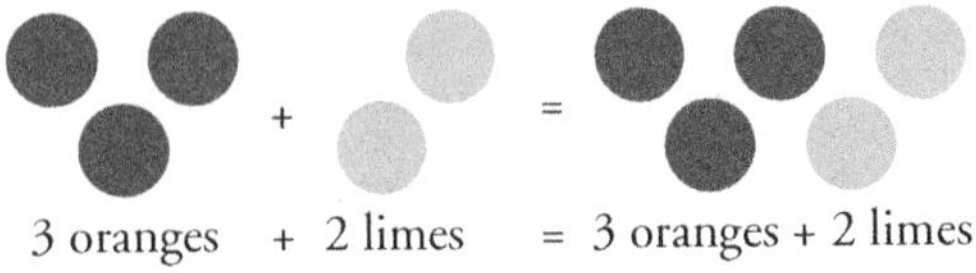

Combining like terms is the general process in which you can simplify a mathematical expression by adding or subtracting the coefficients of like terms.

For example, look at the expression below. Both terms are constants and are, therefore, like terms. You can simplify the value of the expression to 18.

$$8 + 10$$

$$18$$

Consider the following expression:

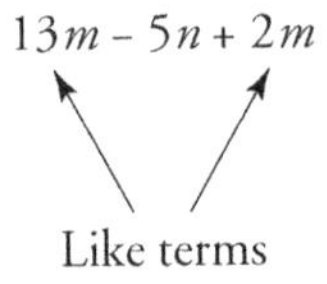

Notice that the term $-5n$ does not share the same variable as the other two terms and, therefore, cannot be combined with the other terms in the expression.

The expression can be simplified to $15m - 5n$.

Example 8:

Simplify $3f - 7g + 8f + 9g$.

Solution:

$11f + 2g$	Combine coefficients of f terms, $3f + 8f = 11f$
	Combine coefficients of g terms, $-7g + 9g = 2g$

Remember, like terms must have the same variable and also the same exponent. For example, x, x^2, and x^3 are NOT like terms.

x 1-dimensional　x^2 2-dimensional　x^3 3-dimensional

Example 9:

Simplify $6x - 2x^2 - 4x + 9x^2$.

Solution:

$7x^2 + 2x$

The x terms cannot be combined with the x^2 terms.

Independent and Dependent Variables

An **equation** is a mathematical statement where two expressions are equivalent.

Mathematical expression	Equation
$4+x$	$4+x=y$

In the equation above, we can refer to x as the **input** value and y as the **output** value because changing the value of x would result in a different value for y.

We can also say that x is the **independent variable** and y is the **dependent variable** because the value of y depends on the value of x.

Example 10:

A clerk at a local ice cream store earns \$10 per hour as shown in the expression $10h = p$, where h represents the number of hours the clerk works in a week and p represents the amount of his weekly paycheck. Identify the independent and dependent variables in this scenario.

Solution:

In this example, h is the *independent variable*, and p is the *dependent variable* because the amount of the clerk's weekly paycheck *depends* on how many hours he works in a week.

Properties of Numbers

Now that you are familiar with the numerical constants and variables we will be working with, let's discuss some of the basic properties of numbers: commutative, associative, distributive, identity, inverse, and properties of zero.

Commutative property

When adding or multiplying, numbers can *commute* or be arranged in a different order and the resulting value will be the same.

Addition	Multiplication
$a+b=b+a$	$ab=ba$
$5+7=12$	$3(6)=18$
$7+5=12$	$6(3)=18$

Note: The same is **<u>not</u>** true of subtraction or division.

Subtraction	Division
$16-9=7$	$\frac{18}{6}=3$
$9-16\neq 7$	$\frac{6}{18}\neq 3$

Associative property

When adding or multiplying, numbers can be *associated* or *grouped* in different arrangements and the resulting value will be the same.

Addition	Multiplication
$a+(b+c)=(a+b)+c$	$a\cdot(b\cdot c)=(a\cdot b)\cdot c$
$2+(3+1)=6$	$4(5\cdot 2)=40$
$(2+3)+1=6$	$(4\cdot 5)2=40$

Note: The same is **not** true of subtraction or division.

Subtraction		Division	
$11-(8+4)$	$(11-8)+4$	$18\div(6+3)$	$(18\div 6)+3$
$11-(12)=-1$	$(3)+4=7$	$18\div(9)=2$	$(3)+3=6$
$11-(8+4)\neq(11-8)+4$		$18\div(6+3)\neq(18\div 6)+3$	

Distributive property

Order of operations dictates that you must simplify expressions inside of grouping symbols first. Consider the mathematical expression below:

$2(1+4)$	
$2(5)$	Combine terms inside parentheses first
10	Multiplication

Now, consider *first* multiplying both terms inside the parentheses by the factor on the outside as such:

$2(1+4)$	
$2\cdot 1+2\cdot 4$	Multiply each term by the outside term
$2+8$	Simplify
10	Addition

Notice that both methods result in the same solution. Therefore, we can generalize the **distributive property** as

$$a(b+c)=a\cdot b+a\cdot c$$

Example 11:

Simplify the expression $9(x-3)$.

Solution:

$9\cdot x-9\cdot 3$	Multiply the inside terms by the outside term
$9x-27$	Simplify

Example 12:

Which expression is equivalent to $-7(y-4)+11(y-3)$?

A. $4y-61$
B. $4y-5$
C. $18y-61$
D. $18y-5$

Solution:

$-7(y)-7(-4)+11(y)+11(-3)$	Multiply the inside terms by the outside terms
$-7y+28+11y-33$	Simplify
$4y-5$	Combine like terms

The answer is Choice B, $4y-5$.

Identity property

Adding 0 or multiplying by 1 preserves the identity of a number:

Addition	Multiplication
$a+0=a$	$a\cdot 1=a$
$9+0=9$	$-7\cdot 1=-7$

Inverse properties

Every real number, *a*, has an **additive inverse**, –*a*, which will *undo* that number and result in a value of 0.

Addition

$$a+(-a)=0$$

$$-17+(17)=0$$

Every real number, except 0, also has a **multiplicative inverse,** $\frac{1}{a}$, which will *undo* that number and result in a value of 1. The product of 0 with any other number is 0, not 1; therefore, zero is the only number without a multiplicative inverse.

Multiplication

$$a \cdot \left(\frac{1}{a}\right) = 1$$

$$4 \cdot \left(\frac{1}{4}\right) = 1$$

Properties of zero

When **adding or subtracting** zero to any number, the value of the original number does not change. We discussed this under "Identity property," but it's worth mentioning again.

Addition Property of Zero

$$a + (0) = a$$

$$9 + 0 = 9$$

When **multiplying** any number by zero, the product is always zero.

Multiplication Property of Zero

$$a \cdot 0 = 0$$

$$\pi \cdot 0 = 0$$

Real Number System

In algebra I, we will be working primarily with the **real number system.** (***Note:*** You will work with imaginary numbers in algebra II.) There are several subsets of numbers in the real number system: natural numbers, whole numbers, integers, rational numbers, and irrational numbers.

Natural numbers—Also referred to as counting numbers: 1, 2, 3, 4, 5, ...

Whole numbers—The set of all natural numbers as well as zero: 0, 1, 2, 3, 4, 5, ...

Integers—The set of all whole numbers as well as negative numbers: ..., –3, –2, –1, 0, 1, 2, 3, ...

Rational numbers—The set of all integers as well as fractions, terminating decimals, or non-terminating decimals with a repeating sequence of numbers: $\frac{1}{2}$, 0.5, $-\frac{1}{3}$, -0.3

Irrational numbers—Any number that *cannot* be expressed as a fraction; this includes non-terminating decimals *without* a repeating sequence of number: $\sqrt{5}$, $\pi = 3.14159265...$, $e = 2.7182...$

Real Number System

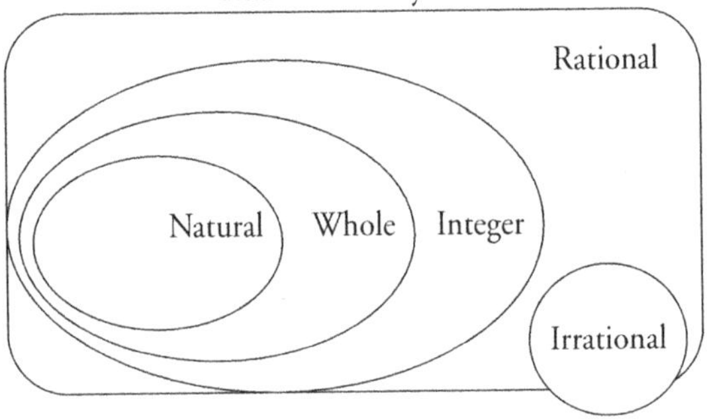

Example 13:

Which of the following real number systems does the number –4 belong to?

I. Natural
II. Whole
III. Integer
IV. Rational
V. Irrational

A. II
B. III
C. III and IV
D. All of the above

Solution:

The answer is Choice C; –4 is an integer and a rational number.

Coordinate Plane

The **coordinate plane** is formed by a horizontal line, the ***x*-axis,** and a vertical line, the ***y*-axis,** which intersect at a point called the **origin.**

The two perpendicular axes split the coordinate plane into four **quadrants.**

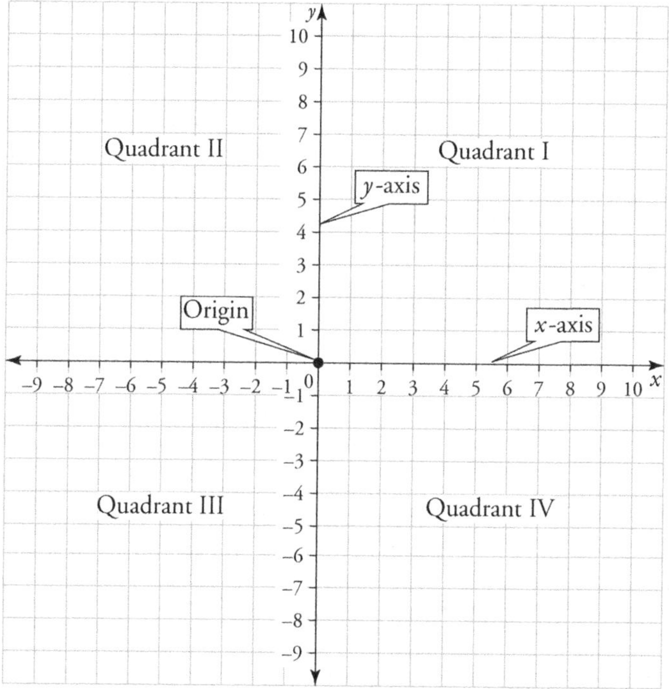

The position of a point on a coordinate plane is identified by two numerical values that describe the point's distance from the origin. This pair of values, referred to as an **ordered pair,** is made up of an ***x*-coordinate** and a ***y*-coordinate,** written as (x, y).

Graphing points in a coordinate plane

Starting at the origin, the *x-coordinate* in an ordered pair (x, y) tells you how many units to move horizontally and in which direction. The *y-coordinate* tells you how many units to move vertically and in which direction.

Example 14:

Plot each point: Q (–2, 3) and P (4, 0).

Solution:

Q (–2, 3): From origin, move 2 units left and 3 units up.

P (4, 0): From origin, move 4 units right and 0 units up or down.

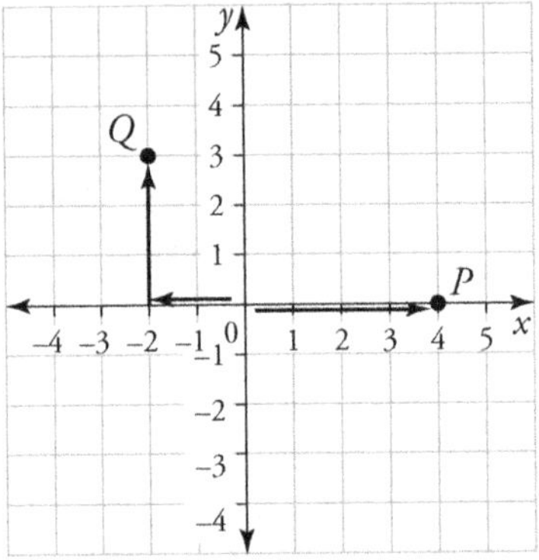

Example 15:

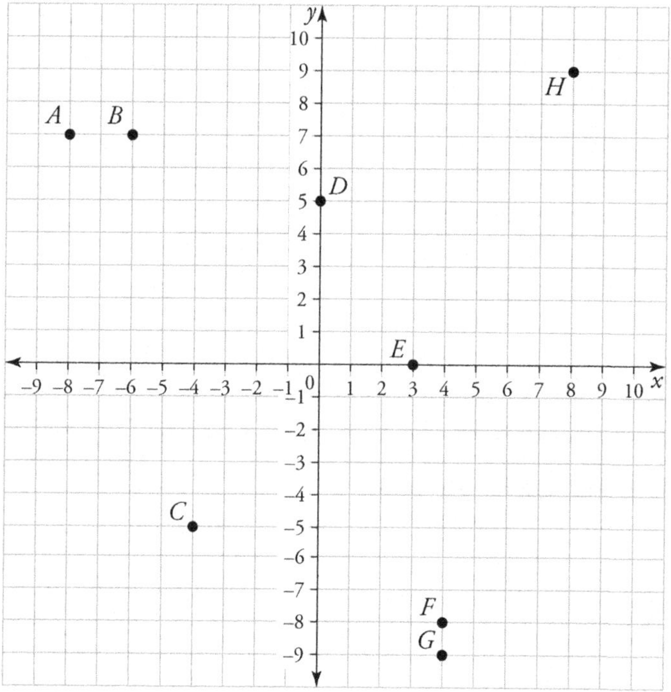

Review the graph above. Match each point *A*–*H* with its corresponding ordered pair below. Name the quadrant in which each point lies.

1. (4, –8)
2. (–4, –5)
3. (0, 5)
4. (3, 0)
5. (4, –9)
6. (8, 9)
7. (–6, 7)
8. (–8, 7)

Solution:

1. (4, –8): *F*, Quadrant IV
2. (–4, –5): *C*, Quadrant III
3. (0, 5): *D*, *y*-axis
4. (3, 0): *E*, *x*-axis
5. (4, –9): *G*, Quadrant IV
6. (8, 9): *H*, Quadrant I
7. (–6, 7): *B*, Quadrant II
8. (–8, 7): *A*, Quadrant II

TECH TIPS: Evaluating and Simplifying Expressions Using a Graphing Calculator

Evaluating expressions

You can evaluate expressions on a graphing calculator by using the table function.

Example:

Evaluate $-3x + 18$ for $x = -6, 10, 25,$ and 180.

Solution:

1. Press [Y=] and enter $-3x + 18$ in Y_1.

Plot 1 Plot 2 Plot 3	
$\backslash Y_1=$	$-3x + 18$
$\backslash Y_2=$	
$\backslash Y_3=$	
$\backslash Y_4=$	
$\backslash Y_5=$	
$\backslash Y_6=$	
$\backslash Y_7=$	

2. Change the table settings in the Table Setup window.

   Press **2nd** **WINDOW**.

   Change the Independent variable from Auto to Ask.

   Note: *You will want to reset your setup back to Auto afterward.*

   Press **ENTER** to save settings.

```
TABLE SETUP
   TblStart = 0
   ΔTbl = 1
Indpnt:  Auto  Ask
Depend:  Auto  Ask
```

3. Press **2nd** **GRAPH** to view the table.

X	Y_1	
-6	36	
10	-12	
25	-57	
180	-522	

X =

4. Enter the values you wish to evaluate for *x*.

   The X column shows the input values you enter.

   The Y_1 column will show the corresponding solutions.

Answer: The solutions are $y = 36, -12, -57$, and -522, respectively.

Simplifying expressions

Consider the following type of question that asks you to simplify an expression:

Example:

Which expression is equivalent to $3(2 + x) - (7 + 4x)$?

A. $x + 11$

B. $8x - 11$

C. $-x - 1$

D. $-x - 7$

Because equivalent expressions will have the same data points, they will also have identical graphs.

Let's compare the tables and graphs for all four expressions to see which answer choice is a simplified version of the original expression.

Solution:

1. Press **Y =** and enter the original expression $3(2 + x) - (7 + 4x)$ in Y_1.
2. Enter the first answer choice into Y_2.

   Press the left arrow key once to highlight the slanted line to the left of Y_2 and change the line settings.
3. Press **ENTER** four times to change the default *thin line* setting to *path style,* which uses a circle to indicate a point as it's being graphed.

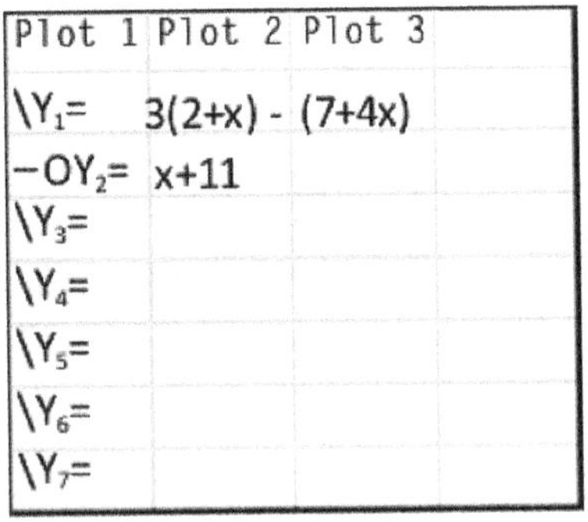

4. Press **GRAPH**.

   Y_1 will graph the original equation.

   Y_2 will graph the first answer choice.

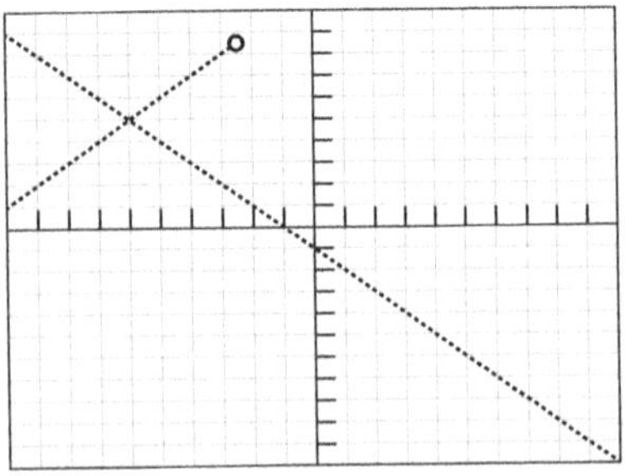

5. If the two expressions are equivalent, Y_2 will trace directly on top of Y_1. Otherwise, you will see two different graphs.
6. Replace the expression in Y_2 with the next answer choice until you find the expression that graphs directly on top of the original expression.

Answer: Choice C has the same graph as the original expression.

Chapter Check-Out

Questions

For questions 1–4, write an algebraic expression for the given phrase.

1. four more than a number *n*
2. eight less than *p*
3. the quotient of *y* and three
4. the product of negative five and *x*

For questions 5–7, write the algebraic expression in words.

5. $9h$
6. $t - 1$
7. $\frac{f}{2}$

For questions 8–12, evaluate each expression for $p = -3$, $q = 6$, and $r = 4$.

8. $5q$
9. $7 - p$
10. $\frac{36}{r}$
11. $8 + 2p$
12. $11 + 2r - q$

For questions 13–15, simplify each expression.

13. $14 - (6)(2)$
14. $(4 + 9)(2 - 7)$
15. $10 - [6 + (5 + 2^2)]$

For questions 16–17, identify the independent and dependent variables.

16. A teacher creates a survey to compare how many hours students study for a test and their test grade.

17. An algebra tutor charges \$40 per hour.

For question 18, simplify the expression, justifying the steps (b and d) and giving the solution (e).

18.

Expression	***Justification***
a. $4(x-2)+7x$	
b. $4(x)-4(2)+7x$	**b.** __________ property
c. $4x-8+7x$	**c.** multiply
d. $4x+7x-8$	**d.** __________ property
e. __________	**e.** combine like terms

For questions 19–20, classify each real number as rational or irrational.

19. $-\frac{2}{7}$

20. $\sqrt{7}$

Answers

1. $n + 4$ (or $4 + n$)
2. $p - 8$
3. $\frac{y}{3}$
4. $-5x$

Questions 5–7 can have multiple responses

5. product of nine and *h*
6. one less than a number, *t*
7. quotient of *f* and two
8. 30
9. 10
10. 9
11. 2
12. 13
13. 2
14. –65
15. –5
16. independent: number of hours the student studies; dependent: student grade
17. independent: number of hours spent tutoring; dependent: total charge
18. **b.** distributive

    d. commutative

    e. $11x - 8$
19. rational
20. irrational

Chapter 2

FUNCTIONS AND EQUATIONS

Chapter Check-In

- ❑ Functions
- ❑ Function notation
- ❑ Domain and range
- ❑ Mapping diagrams
- ❑ Vertical-line test
- ❑ Sequences
- ❑ Writing equations
- ❑ TECH TIPS: Generating a sequence on a graphing calculator

Now that we've familiarized ourselves with the language of algebra, let's explore how the basic structures of algebra work.

Functions

Picture a machine like the one below into which you can input any number, x. The machine performs a set of predetermined mathematical operations and outputs a solution.

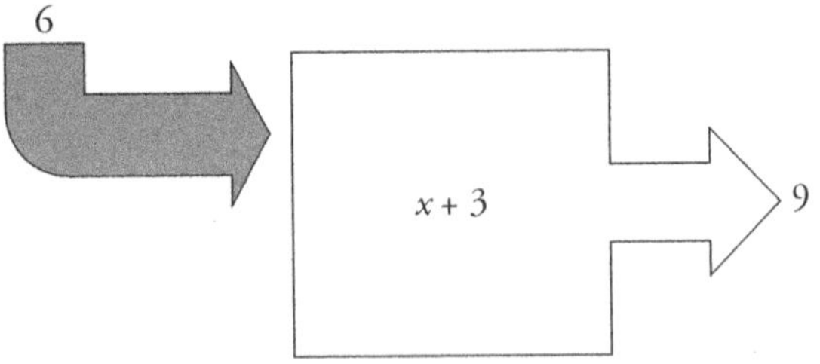

When the machine functions correctly, it takes any input value and then adds 3.

When the input value is 6, the output value is 9. There could be no other possible output when an input value of 6 is used in this machine.

If a value other than 9 was the output in this scenario, then one could say that the machine no longer functions.

A **function** is a mathematical relationship in which every input has *exactly one* output value.

We can use the equation $y = x + 3$ to represent the function in the function machine shown above, where x represents the independent variable, or input, and y represents the dependent variable, or output.

We can say that the value of this function depends on x.

Function Notation

In **function notation,** we can represent an equation as a function of x.

Equation	Function notation
$y = x + 3$	$f(x) = x + 3$

When we evaluate the function $f(x) = x + 3$ for $x = 6$, the solution is 9. You can express this in function notation with the following statement:

$$f(x) = x + 3$$
$$f(6) = 6 + 3$$
$$f(6) = 9$$

(***Note***: In function notation, $f(x)$ is interchangeable with y.)

Example 1:

Evaluate $f(x) = 2x + 8$ for $x = -3$.

Solution:

$f(-3) = 2(-3) + 8$	Substitute -3 for x
$f(-3) = -6 + 8$	Multiplication
$f(-3) = 2$	Addition

Example 2:

Evaluate $f(x)=\frac{x^2}{4}$ for $x = 8$.

Solution:

$$f(8)=\frac{(8)^2}{4} \quad \text{Substitute 8 for } x$$

$$f(8)=\frac{64}{4} \quad \text{Simplify exponent}$$

$$f(8)=16 \quad \text{Division}$$

Domain and Range

Another way of referring to input and output values is using the terms *domain* and *range.* **Domain** refers to the set of all input values in a function. **Range** refers to the set of all output values in a function.

Here are some other terms that refer to x: *independent variable, input value,* and *domain.* Some terms that refer to y are *dependent variable, output value,* and *range.*

Let's evaluate a function for more than one value in the domain.

Example 3:

Find the range (*output values*) of the following function with a domain of x: {–2, –1, 0, 1, 2}

$$f(x) = 3x$$

Solution:

$$f(-2)=3(-2)=-6$$
$$f(-1)=3(-1)=-3$$
$$f(0)=3(0)=0$$
$$f(1)=3(1)=3$$
$$f(2)=3(2)=6$$

The range of the function with a domain x: {–2, –1, 0, 1, 2} is $f(x)$: {–6, –3, 0, 3, 6}.

We can also arrange the input and output values in a table like the one below:

x	–2	–1	0	1	2
f(x)	–6	–3	0	3	6

Mapping Diagrams

You can also use a mapping diagram to organize the input and output values of a function. **Mapping diagrams** match the elements in the domain with their corresponding elements in the range. Mapping diagrams are also helpful in determining whether a mathematical expression represents a function. Here is a mapping diagram of the function $f(x) = 3x$.

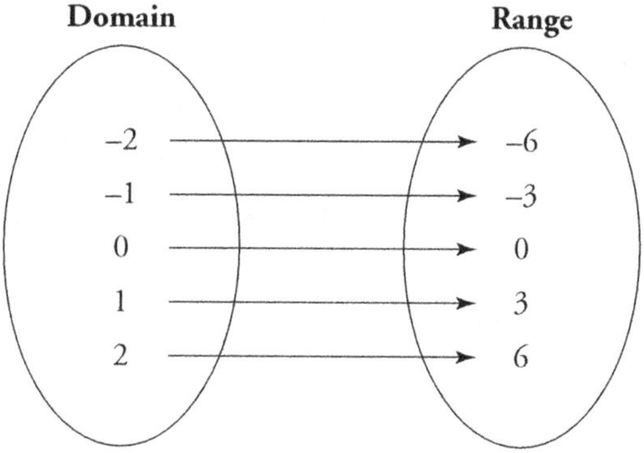

Example 4:

Use the mapping diagram to state the domain and range of the relation and state whether it is a function.

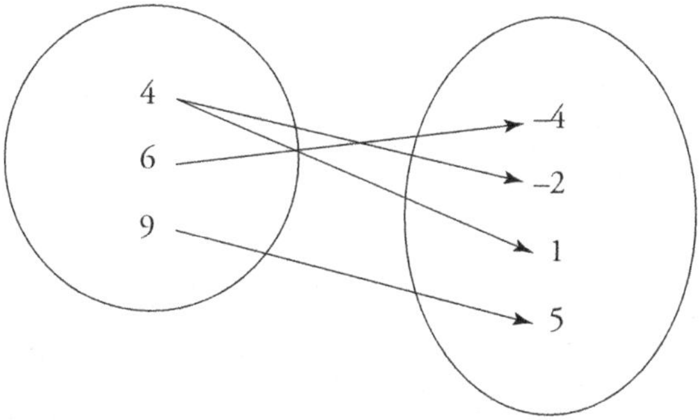

Solution:

Domain: {4, 6, 9}

Range: {–4, –2, 1, 5}

This relation is NOT a function because the domain value 4 has more than one output value.

Vertical-Line Test

Another method for determining whether a mathematical relation is a function is using the **vertical-line test.**

We know that a relation is a function if each input value has *exactly* one output value.

Consider a relation that contains the points (3, 1) and (3, 5). This relation is NOT a function because the input value, 3, is assigned two different output values, 1 and 5.

When the two points are plotted on a coordinate grid, notice that drawing a vertical line through 3 on the *x*-axis would intersect *both* points.

Vertical-Line Test

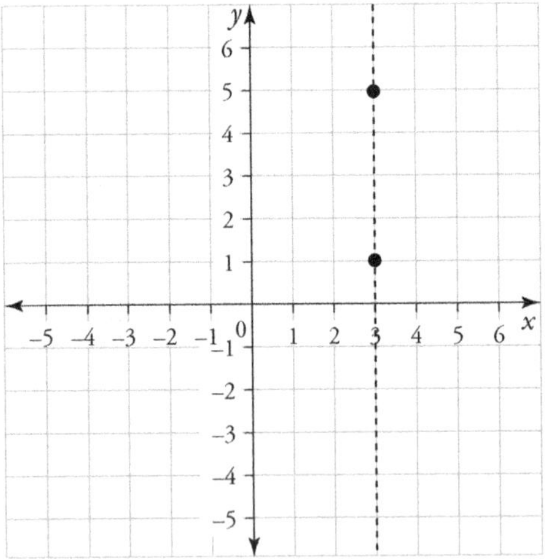

> **Vertical-Line Test:**
>
> Whenever a vertical line can intersect the graph of a relation at more than one point, we can say that the relation fails the vertical-line test and, therefore, is NOT a function.

Example 5:

Use the vertical-line test to determine if each relation in the following graphs is a function.

Graph 1

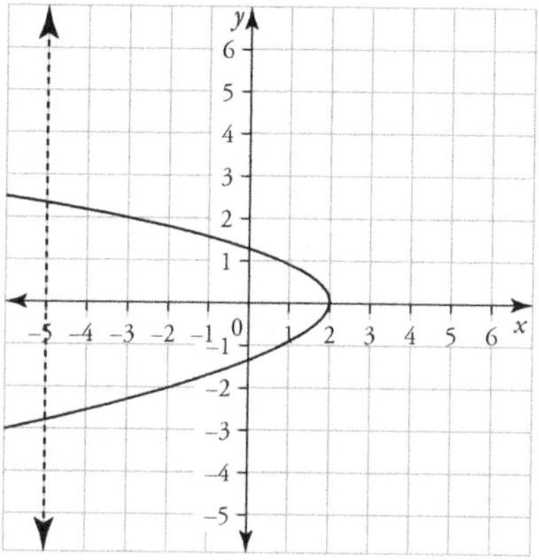

Graph 2

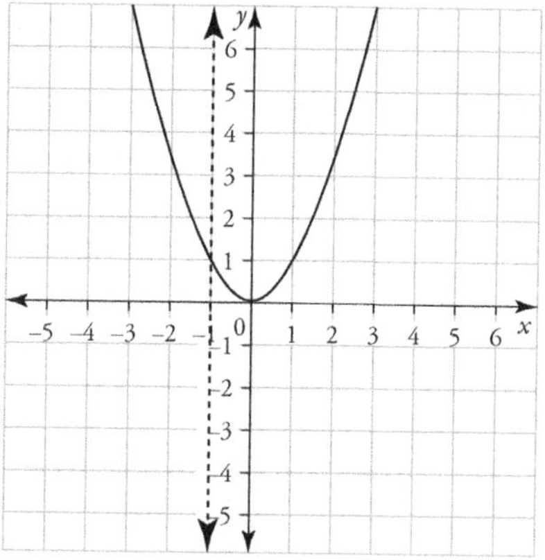

Graph 3

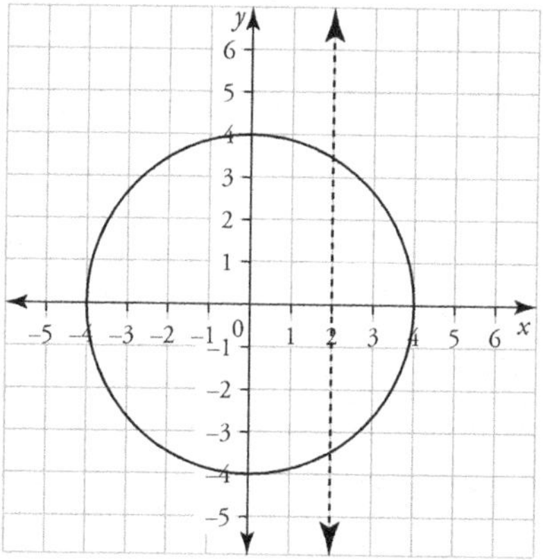

Solution:

Graph 2 passes the vertical-line test; we can conclude that it represents a function because it intersects the graph of the relation at only one point. Graph 1 and Graph 3 do not pass the vertical-line test; those relations are not functions.

Sequences

A **sequence** is a list of numerical terms. The list, or sequence, is made up of values from the range.

$$13, 10, 7, 4, \ldots$$

The numbers in the sequence are called **terms.** The variable a represents the terms in a sequence, and n is a whole number that represents the term's position in the sequence.

a_n nth term
a_1 1st term
a_2 2nd term

Note: The term's position in the sequence is written as a subscript of a.

Term's position	1	2	3	4...	n
	↓	↓	↓	↓	
Term	**13,**	**10,**	**7,**	**4...**	
	a_1	a_2	a_3	a_4	a_n

Since a_n represents any number in a sequence, we can also use

a_{n-1} to represent the previous number in a sequence, and
a_{n+1} to represent the next number in a sequence

Arithmetic sequences

Let's take a look at a table of values where instead of x and y, the input and output values are denoted by n and a_n, respectively.

n	1	2	3	4	5
a_n	8	5	2	−1	−4

If n represents the position of a term in a sequence, then the following are true:

The 1st term is 8.

The 2nd term is 5.

The 3rd term is 2.

The 4th term is –1.
The 5th term is –4.

The sequence would look like this:

$$8, 5, 2, -1, -4, \ldots$$

If we take a look at the terms in the sequence, you might see that the values are decreasing by 3.

```
8,   5,   2,  –1,  –4, ...
 \  / \  / \  / \  /
  –3   –3   –3   –3
```

The **common difference** is the difference between any two consecutive terms.

$5-8=-3 \qquad 2-5=-3 \qquad -1-2=-3 \qquad -4-(-1)=-3$

A sequence where consecutive terms have a common difference is referred to as an **arithmetic sequence.**

Geometric sequences

Now, consider the sequence below:

$$a_n = 2, 8, 32, 128, 512, \ldots$$

Find the differences between the terms in the sequence:

```
2,   8,  32, 128, 512
 \  / \  / \  / \  /
  6   24   96  384
```

As you can see, the differences between consecutive terms in the sequence are *not* common.

However, you may have noticed that the terms in the sequence are being multiplied by a factor of 4 to get the next term in the sequence.

```
2,   8,  32, 128, 512
 \  / \  / \  / \  /
  ×4   ×4   ×4   ×4
```

Dividing any element in the sequence by the previous number in the sequence reveals a **common ratio.**

$$\frac{8}{2}=4 \qquad \frac{32}{8}=4 \qquad \frac{128}{32}=4 \qquad \frac{512}{128}=4$$

A sequence with a common ratio of any term to its previous term is referred to as a **geometric sequence.**

Example 6:

Determine if the following sequence is arithmetic or geometric. If it is arithmetic, find the common difference. If it is geometric, find the common ratio.

$$35, 28, 21, 14, \ldots$$

Solution:

The sequence is arithmetic because there is a common difference of –7.

Example 7:

Determine if the following sequence is arithmetic or geometric. If it is arithmetic, find the common difference. If it is geometric, find the common ratio.

$$81, 27, 9, 3, \ldots$$

Solution:

The sequence is geometric because there is a common ratio of $\frac{1}{3}$.

Finding the *n*th term of a sequence

To find the indicated term of a sequence, substitute the position of the term in the sequence and simplify.

Example 8:

Find the indicated term in the sequence.

9th term: $a_n = -8n + 59$

$a_9 = -8(9) + 59$	Substitute 9 for n
$a_9 = -72 + 59$	Multiply
$a_9 = -13$	Add

Example 9:

Find the next three terms in each sequence.

A. –8, –5, –2, 1, …

B. 2, 12, 72, 432, …

Solution:

A. –8, –5, –2, 1, ...

\ / \ / \ /

+3 +3 +3	The common difference is +3
$1 + 3 = 4$	Add 3 to the 4th term to find the 5th term
$4 + 3 = 7$	Add 3 to the 5th term to find the 6th term
$7 + 3 = 10$	Add 3 to the 6th term to find the 7th term

B. 2, 12, 72, 432, ...

\ / \ / \ /

×6 ×6 ×6	The common ratio is 6
$432 \cdot (6) = 2{,}592$	Multiply the 4th term by 6
$2{,}592 \cdot (6) = 15{,}552$	Multiply the 5th term by 6
$15{,}552 \cdot (6) = 93{,}312$	Multiply the 6th term by 6

Writing Equations

Now let's focus on how to write a function rule, or an **equation,** to represent a table of values or a sequence.

Consider again the sequence in Example 6.

35, 28, 21, 14, ...

In order to produce the sequence of numbers, our function machine might look like this:

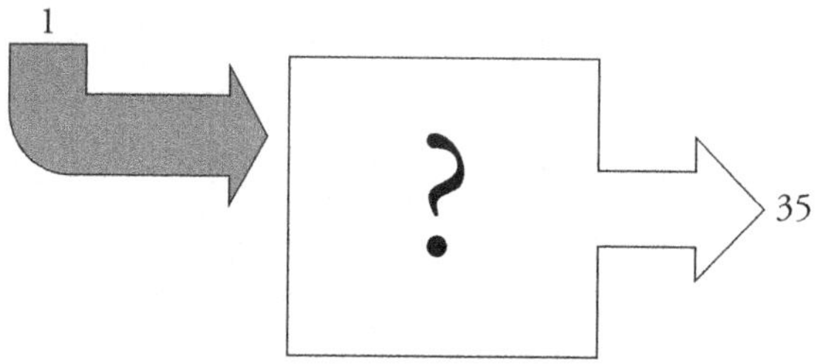

The first element in the sequence is 35, but the function rule is still unknown. There are a couple of operations that could turn a 1 into a 35.

Adding 34	Multiplying by 35
$1 + 34 = 35$	$(35)1 = 35$

If we write an equation where x represents the input and y represents the output, we would have:

Adding 34	Multiplying by 35
$x + 34 = y$	$35x = y$

It appears as if both of these equations would work in this scenario. Both function rules take an input value of 1 and result in an output value of 35.

Now let's consider the other values in our sequence. Create a table of values where x represents the position of the term in the sequence and y represents the term itself:

x	*y*
1	35
2	28
3	21
4	14

Now apply the same function rule for the first ordered pair to the rest of the input values.

$x + 34 = y$	$35x = y$
$(2) + 34 \neq 28$	$35(2) \neq 28$
$(3) + 34 \neq 21$	$35(3) \neq 21$
$(4) + 34 \neq 14$	$35(4) \neq 14$
$(5) + 34 \neq 7$	$35(5) \neq 7$

Notice that these function rules do NOT work for any of the other ordered pairs.

We should now consider other possible function rules that would take an input value of 1 and yield an output value of 35.

Think of rules that use more than one mathematical operation. For example, multiplying 1 by 20 and then adding 15 could also result in 35.

$$20(1)+15=35$$
$$20x+15=y$$

Again, let's test this possible function rule on the remainder of our ordered pairs:

$$20(2)+15=55$$
$$20(2)+15\neq 28$$

We see immediately that this function rule also does not work for every ordered pair in our sequence.

Ultimately, there are infinitely many possible function rules that could turn an input of 1 into an output of 35.

Now let's consider some more information that could make the process of finding the function rule easier.

Using the common difference to write an equation

Earlier we identified that there was a common difference of –7 between consecutive terms in the sequence. Another way of stating this is by saying that *to find the next term in the sequence, we take the previous term and subtract 7.*

To find the second term in the sequence, we would have to subtract 7 only *once.*

$$35-7=28$$

To find the third term in the sequence, we would have to subtract 7 a *second* time.

$$35-7-7=21$$

This pattern demonstrates that we need to multiply by our common difference of –7 every time to find the next value. We can represent that symbolically with the expression

$$-7x$$

We can write the equation

$$y = -7x$$

However, you can quickly see that multiplying the input values by –7 doesn't yield the correct output values. A second operation must be needed.

x	**–7x**	**y**
1	–7(1)= –7	35
2	–7(2)= –14	28
3	–7(3) = –21	21
4	–7(4) = –28	14
5	–7(5) = –35	7

If we consider a second operation to reach the desired output, we can see that adding 42 after multiplying each input value by –7 yields the correct output.

x	**–7x**	**y**
1	–7(1) = –7 + 42 = 35	35
2	–7(2) = –14 + 42 = 28	28
3	–7(3) = –21 + 42 = 21	21
4	–7(4) = –28 + 42 = 14	14
5	–7(5) = –35 + 42 = 7	7

Therefore, the function rule, or equation that gives us the sequence 35, 28, 21, 14, ... is $y = -7x + 42$.

> **Critical Connection:**
>
> Notice that the **common difference** (the rate at which the values changed) becomes the *coefficient* of the independent variable, x, in the equation.
>
> If you use that common difference to count *backward from the first term* to the previous term or the zero term, the output is 42. Notice that the value of that zero term becomes the *constant* in the equation.

Using the zero term to write an equation

Finding the Zero Term

"Zero" term	1st term	2nd term	3rd term	4th term
42	35,	28,	21,	14...

+7 -7 -7 -7

$y = -7x + 42$

Common difference → -7; Zero term → 42

Use the common difference and count backward from the first term to find the zero term.

Example 10:

Write an equation for the following table of values.

x	0	1	2	3	4
y	–5	–3	–1	1	3

Solution:

There is a common difference of 2, which represents the coefficient of x.

The zero term is –5, which represents the constant in the equation.

The output of the equation is found by multiplying the input value by 2 and then subtracting 5. Symbolically, we would write this as follows:

$$y = 2x - 5$$

Example 11:

The relationship in the table below represents a function. If h represents the number of hours the plumber works, and f represents the fee he charges, write an equation that represents this relationship.

Number of Hours a Plumber Works	Fee in Dollars
1	105
2	140
3	175
4	210

Solution:

The common difference is increasing by $35 per hour.

There is an initial charge of $70; this is the zero term. You can find the zero term by subtracting the common difference, 35, from the first term, 105.

$$f = 35h + 70$$

TECH TIPS: Generating a Sequence on a Graphing Calculator

Example:

Find the first five terms of the sequence $a_n = 7n - 3$.

Solution:

1. From the home screen, press **2nd** **STAT** to access the "LIST" menu.

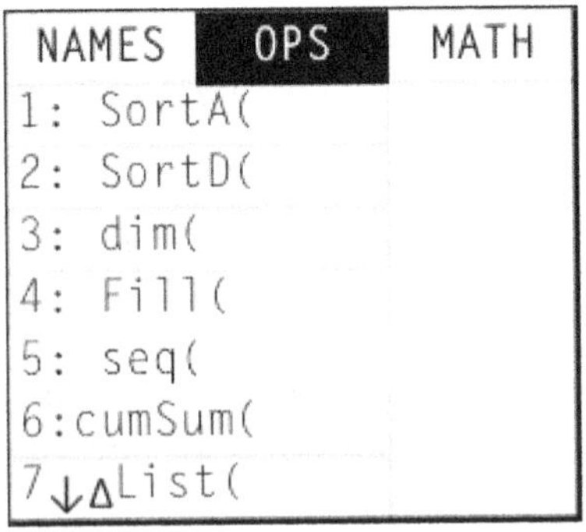

2. Press the right arrow key to highlight the OPS menu.

3. Arrow down to highlight `5: seq` (or press [5].

   You will be taken back to the home screen.

4. Enter the sequence rule, $7x - 3$.

   Press [,] and press [x,T,θ,π]

   to identify the variable you are using, *x*.

   Press [,] and enter the first term you want to generate, 1.

   Press [,] and enter the last term you want to generate, 5.

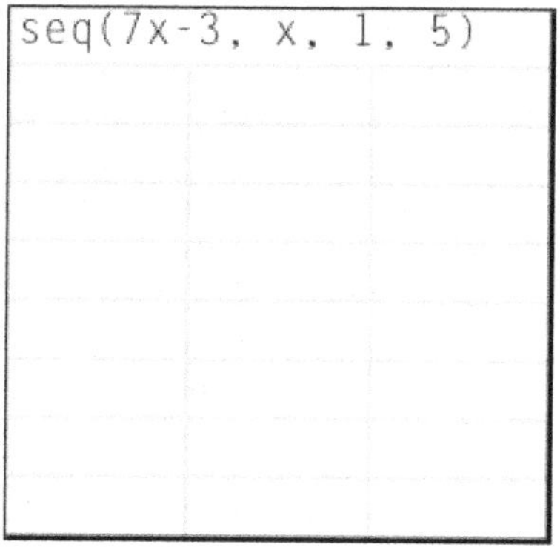

5. Press [ENTER] to generate the first five terms of the sequence.

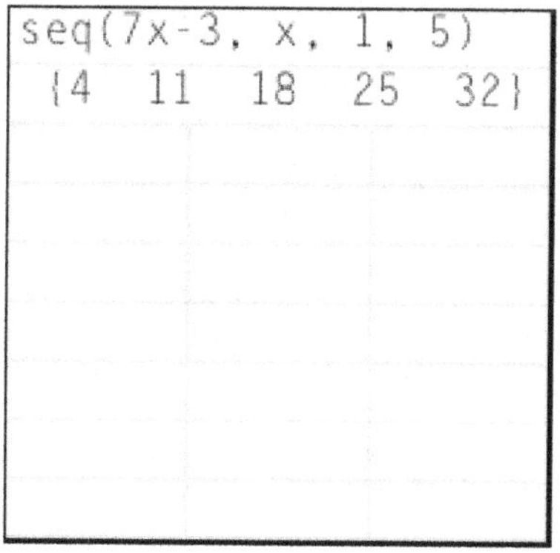

Answer: 4, 11, 18, 25, 32

Example:

Generate the next four terms in the sequence 11, 7, 3, –1, ...

Solution:

1. Find the equation for the sequence.

   11, 7, 3, –1, ...

   \ /\ / \ /

   –4 –4 –4 — The common difference is –4

   $11 + 4 = 15$ — Working backward, the zero term is 15

   $a_n = -4n + 15$ — Write the common difference, –4, as the coefficient and the zero term, 15, as the constant

2. Follow steps 1–4 from the first example.
3. Press [,] and press [x, T, θ, π] to identify the variable, *x*.
   Press [,] and enter the first term you want to generate, 5 for the 5th term.
   Press [,] and enter the last term you want to generate, 8 for the 8th term.

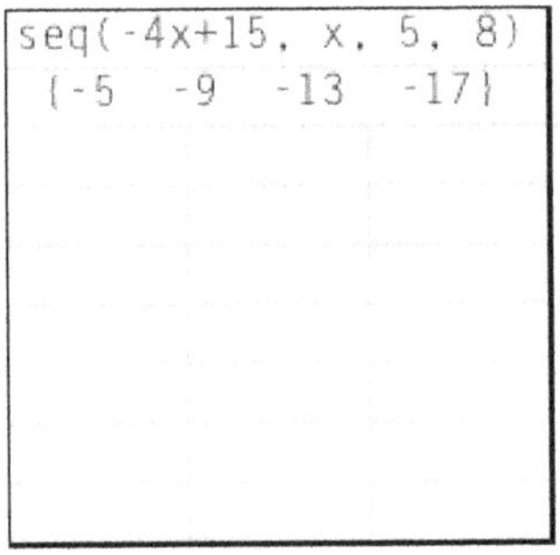

4. Press [ENTER] to generate the next four terms of the sequence, terms 5–8.

Answer: The next four terms in the sequence are –5, –9, –13, –17.

Chapter Check-Out

Questions

For questions 1–2, evaluate each function.

1. $g(x) = 3x - 5$, when $x = 1$ and when $x = -2$

2. $h(x) = \frac{2}{5}x + 4$, when $x = 15$ and when $x = -5$

For questions 3–4, give the domain and range of each relation. Determine whether or not the relation is a function.

3.

x	–2	–1	0	1	2
y	3	0	–1	0	3

4. {(2, 2), (1, 1), (0, 0), (1, –1), (2, –2)}

5. Which of the following does NOT represent the function $f(x) = 4x - 6$?

A. {(–4, –22), (2, 2), (4, 10), (5, 14)}

B.

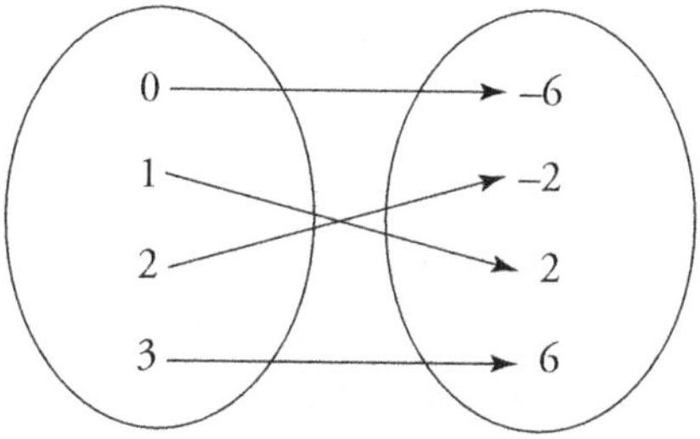

C.

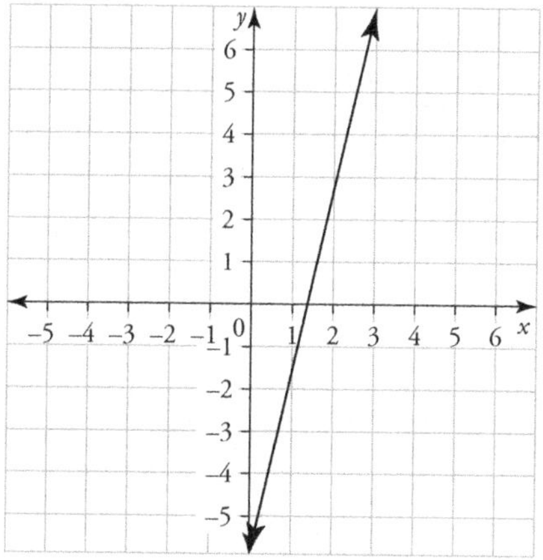

D.

x	*f(x)*
–3	–18
0	–6
1	–2
3	6

For questions 6–7, determine if the sequence is arithmetic or geometric. If it is arithmetic, find the common difference. If it is geometric, find the common ratio.

6. 2, 6, 18, 54, ...

7. 13, 10, 7, 4, ...

For questions 8–9, find the *n*th term of the sequence.

8. 12th term; $a_n = -7n + 5$

9. 8th term; $a_n = 3(2)^{n-1}$

For questions 10–11, find the next four terms in the sequence.

10. 6, 1, –4, –9, ...

11. –4, 12, –36, 108, ...

For questions 12–13, write an equation that describes the relationship shown.

12. Jeff purchases an Internet gaming subscription that charges an initial sign-up fee of $50, plus an additional $10 per month.

13.

Number of minutes air conditioner is on	5	10	15	20
Temperature in room (°F)	86	83	80	77

Answers

1. $-2; -11$
2. 10; 2
3. domain: $\{-2, -1, 0, 1, 2\}$;
   range: $\{-1, 0, 3\}$;
   the relation is a function
4. domain: $\{0, 1, 2\}$;
   range: $\{-2, -1, 0, 1, 2\}$;
   the relation is NOT a function
5. B
6. geometric;
   common ratio = 3
7. arithmetic;
   common difference = -3
8. -79
9. 384
10. $-14, -19, -24, -29$
11. $-324, 972, -2916, 8748$
12. $f(x) = 10x + 50$
13. $f(x) = \frac{-3}{5}x + 89$

Chapter 3

SOLVING EQUATIONS

Chapter Check-In

- ❑ Properties of equality
- ❑ One-step equations
- ❑ Multi-step equations
- ❑ Equations with variables on both sides
- ❑ Literal equations
- ❑ Different types of solutions for linear equations
- ❑ TECH TIPS: Solving equations with a graphing calculator

Properties of Equality

Let's take a deeper look at the general structure of an **equation.** An equation consists of two mathematical expressions that are equivalent, denoted by the equal sign.

It is often helpful to picture an equation as being a two-sided scale where both sides are perfectly balanced, as shown in the diagram below.

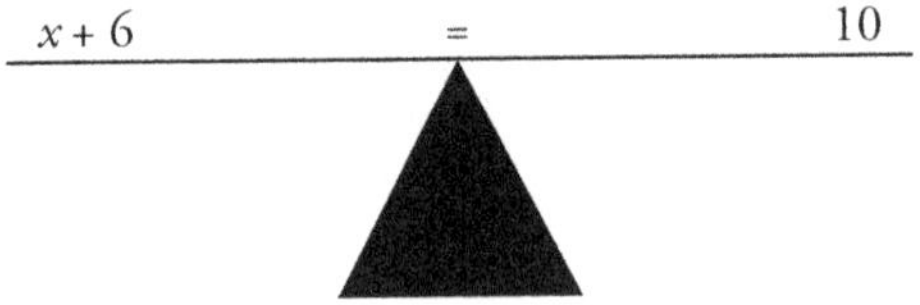

Consider what would happen if you suddenly increased the value of only one side of the scale.

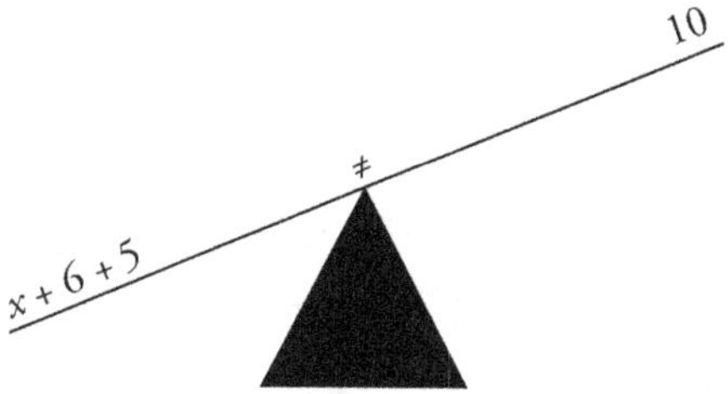

The scale would instantly tip in the direction of the heavier side. However, if you added the same amount to BOTH expressions on either side of the equal sign, then they would remain perfectly balanced.

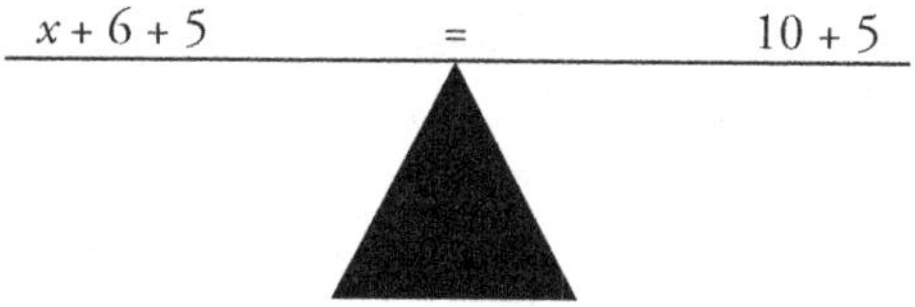

This elementary mathematical concept is the basis for the **properties of equality,** which state that two expressions in an equation will remain equal as long as the same operations are performed on both sides.

Addition

If $a = b$, then $a + c = b + c$.

$$2 + 4 = 6$$
$$6 = 6$$
$$2 + 4 + 1 = 6 + 1$$
$$7 = 7$$

Subtraction

If $a = b$, then $a - c = b - c$.

$$3 + 5 = 8$$
$$8 = 8$$
$$3 + 5 - 7 = 8 - 7$$
$$1 = 1$$

Multiplication	Division
If $a = b$, then $ac = bc$.	If $a = b$, then $\frac{a}{c} = \frac{b}{c}$; $c \neq 0$.
$4+1=5$	$5+7=12$
$5=5$	$12=12$
$(4+1)2=(5)2$	$\frac{5+7}{3}=\frac{12}{3}$
$10=10$	$4=4$

These properties are collectively sometimes referred to as the **Golden Rule of Algebra.**

Golden Rule of Algebra:

"What you do to one side of an equation, you must do to the other."

Example 1:

Fill in the missing expression that would make the equation in Line 1 also true in Line 2.

Line 1: $t = 9$

Line 2: $\boxed{?} = 2$

A. $\boxed{t+2} = 2$
B. $\boxed{t-2} = 2$
C. $\boxed{t-7} = 2$
D. $\boxed{t+7} = 2$

Solution:

The answer is C. The expression on the right side of the equal sign is reduced by 7; therefore, both sides of the equation will remain equal if the expression on the left side of the equal sign, *t*, is also reduced by 7.

One-Step Equations

Addition and subtraction

Imagine that you and a friend decide to put your money together to purchase a pizza. The cost of a pizza at the local pizzeria is $12. If you write an equation to represent this situation, where m represents the amount of money you contribute to the purchase of the pizza, and f represents the amount of money your friend pays, the equation would look like this:

$$m + f = 12$$

If your friend has $5 to contribute, how much money would you have to pay in order to have exactly enough money to purchase the pizza?

You can substitute 5 into the equation for the variable f:

$$m + 5 = 12$$

To answer the question, you must now solve the equation for m.

Recall that in Chapter 1 we discussed inverse properties, which outline how you can basically "undo" any number (see pages 14–15).

Inverse Property of Addition:

$$a + (-a) = 0$$

When trying to solve for a variable, you must "undo" every value on the same side of the equal sign in order to get the desired variable by itself. We call this process **isolating the variable.**

Adding (–5), or subtracting 5, will undo the number 5 in $m + 5 = 12$. Remember that in order for the equation to remain balanced, you must perform the same operation on both sides.

$$\begin{array}{rcl} m + 5 &=& 12 \\ -5 && -5 \\ \hline m &=& 7 \end{array} \quad \text{Subtract 5 from both sides}$$

Check for accuracy:

To check your solution, substitute the value back into the equation and evaluate.

$$(7) + 5 = 12 \checkmark$$

Therefore, if your friend pays $5, you must pay $7 to purchase a $12 pizza.

Here are some more examples.

Example 2:

Solve for p: $13 + p = 42$.

Solution:

$$\begin{array}{rcl} 13 + p &=& 42 \\ -13 & & -13 \quad \text{Additive inverse} \\ \hline p &=& 29 \end{array}$$

Example 3:

Solve for r: $r - 6.2 = 4.7$.

Solution:

$$\begin{array}{rcl} r - 6.2 &=& 4.7 \\ +6.2 & & +6.2 \quad \text{Additive inverse} \\ \hline r &=& 10.9 \end{array}$$

Multiplication and division

The previous examples showed how you can justify isolating a variable using the inverse properties of addition and subtraction. You can also use division to undo multiplication equations, and vice versa.

Recall the Inverse Property of Multiplication:

$$a \cdot \left(\frac{1}{a}\right) = 1;\ a \neq 0$$

Example 4:

Jimmy earns \$8.50 per hour as a dog groomer. He wants to save up for a video game that costs \$59.50. How many hours, h, must Jimmy work in order to have enough money from his paycheck, p, for the video game?

Solution:

$8.5h = p$

$8.5h = 59.5$ Substitute 59.50 for p

$\frac{8.5h}{8.5} = \frac{59.5}{8.5}$ Multiplicative inverse; divide both sides by 8.5

$h = 7$ Simplify

Jimmy must work 7 hours to earn enough to purchase the video game.

Example 5:

Solve for x: $\frac{x}{9} = 4.2$.

Solution:

$\frac{x}{9} = 4.2$ x is being divided by 9

$(9)\frac{x}{9} = 4.2(9)$ Multiplicative inverse; multiply both sides by 9

$x = 37.8$ Simplify

Check for accuracy: $\frac{37.8}{9} = 4.2$ ✓

Critical Connection:

Dividing by a number is the same as multiplying by the reciprocal. For example, dividing by 2 is the same as multiplying by $\frac{1}{2}$.

When an equation contains fractions, it can be easier to multiply by the reciprocal instead of dividing.

$$\frac{4}{3}g = 24 \qquad g \text{ is being multiplied by } \frac{4}{3}$$

$$\left(\frac{3}{4}\right)\frac{4}{3}g = 24\left(\frac{3}{4}\right) \qquad \text{The reciprocal of } \frac{4}{3} \text{ is } \frac{3}{4}$$

$$g = 18$$

Example 6:

Solve for v: $\frac{2}{5}v = \frac{3}{7}$.

Solution:

$$\frac{2}{5}v = \frac{3}{7} \qquad v \text{ is being multiplied by } \frac{2}{5}$$

$$\left(\frac{5}{2}\right)\frac{2}{5}v = \frac{3}{7}\left(\frac{5}{2}\right) \qquad \text{The reciprocal of } \frac{2}{5} \text{ is } \frac{5}{2}$$

$$v = \frac{15}{14}$$

Multi-Step Equations

In Chapter 1, we saw equations that contained more than one operation such as addition and multiplication like the example below.

Example 7:

A professional caterer charges $7 per person and an initial fee of $50 for catering your school club luncheon. You have a food budget of $190. How many guests can you invite to your luncheon?

Solution:

$$7g + 50 = 190$$

Price per guest — Initial fee — Budget

When evaluating an expression with more than one operation, you must follow order of operations (see page 8 in Chapter 1). However, since you are trying to "undo" these operations to solve for a variable, you must now work backward.

$7g+50=190$	First, g is being multiplied by 7, then 50 is being added
$-50 \quad -50$	Working backward, subtract 50 from both sides
$7g \quad =140$	Simplify
$\frac{7g}{7}=\frac{140}{7}$	Since g is being multiplied by 7, divide both sides by 7
$g=20$	

You can invite 20 guests to your school club luncheon.

Some equations need to be simplified first before solving. Often, you may be able to use the distributive property and/or combine like terms before beginning to isolate the variable.

Example 8:

Solve for x: $2(x-3)+4x=72$.

Solution:

$2(x-3)+4x=72$	Distribute 2 into the parentheses, multiply by 2
$2x-6+4x=72$	Simplify
$6x-6=72$	Combine like terms
$+6 \quad +6$	Working backward, add 6 to both sides
$6x=78$	Simplify
$\frac{6x}{6}=\frac{78}{6}$	Divide both sides by 6
$x=13$	Simplify

Equations with Variables on Both Sides

Sometimes, equations will have variables on both sides of the equal sign. This is often used when comparing two different expressions. To solve equations with variables on both sides, first move all the variables to one side of the equal sign.

Example 9:

Your school club is raising funds for your class trip. You have decided to sell t-shirts with the school logo. One manufacturer charges $6 per t-shirt plus a $30 shipping fee. A second manufacturer charges $7.50 per t-shirt plus a $15 shipping fee. Which manufacturer offers the better deal?

Solution:

You can compare both expressions in an equation to find which manufacturer offers the better deal. Let s represent the number of t-shirts purchased and c represent the total cost.

Manufacturer #1	**Manufacturer #2**
$6s+30=c$	$7.5s+15=c$

When comparing the two expressions, write it as one equation:

$$6s + 30 = 7.5s + 15$$

Move all the variables to one side of the equal sign using their additive inverse.

$\begin{array}{r} 6s+30=7.5s+15 \\ -6s \qquad\quad -6s \\ \hline \end{array}$	Undo $6s$ on the left by subtracting $6s$ from both sides
$30=1.5s+15$	Simplify $7.5s-6s$
$\begin{array}{r} -15 \qquad\quad -15 \\ \hline \end{array}$	Work backward, subtract 15 from both sides
$15=1.5s$	Simplify
$\dfrac{15}{1.5}=\dfrac{1.5s}{1.5}$	Divide by 1.5
$10=s$	

When you purchase 10 t-shirts from either manufacturer, the cost will be the same. You can use this to decide which manufacturer to use depending on if you plan on selling more than or fewer than 10 t-shirts.

Example 10:

Solve for x: $3(x + 9) = 8x + 21$.

Solution:

$3(x+9)=8x+21$	
$3x+27=8x+21$	Distributive property
$-3x \quad -3x$	Additive inverse; collect variables on one side
$27=5x+21$	Simplify
$-21= \quad -21$	Additive inverse; subtract 21 from both sides
$6=5x$	Simplify
$\frac{6}{5}=\frac{5x}{5}$	Multiplicative inverse; divide both sides by 5
$1.2=x$	Simplify

Literal Equations

Literal equations are formulas that show the relationship between different values.

For example, a **rate** can be calculated as a ratio of distance, d, covered over a specific amount of time, t, often expressed as *miles per hour* or *feet per second.*

$$r=\frac{d}{t}$$

This specific formula is convenient when the distance covered and time elapsed are given and you are asked to solve for the rate.

However, when you are asked to solve for distance, d, or time, t, it would be more convenient to rearrange the formula or solve the literal equation for the desired variable.

Example 11:

Solve the formula $r=\frac{d}{t}$ for d.

Solution:

$$(t)r = \frac{d}{t}(t) \quad \text{Multiplicative inverse; multiply both sides by } t$$

$$rt = d$$

The resulting formula indicates that distance is the product of rate and time.

Example 12:

Solve the formula $r = \frac{d}{t}$ for t.

Solution:

$$(t)r = \frac{d}{t}(t) \quad \text{Multiplicative inverse; multiply both sides by } t$$

$$rt = d \quad \text{Simplify}$$

$$\frac{rt}{r} = \frac{d}{r} \quad \text{Multiplicative inverse; divide both sides by } r$$

$$t = \frac{d}{r}$$

The resulting formula indicates that the time required to travel a certain distance is equal to the ratio of distance traveled divided by the rate at which the object is traveling.

By rearranging the variables, you can transform one formula into three different literal equations, each one useful for solving for a different value.

$$d = rt \qquad t = \frac{d}{r} \qquad r = \frac{d}{t}$$

Example 13:

The formula for converting temperature in degrees Fahrenheit to degrees Celsius is $C = \frac{5}{9}(F - 32)$. What is the formula for converting temperature in degrees Celsius to degrees Fahrenheit? (***Hint***: Solve for F.)

Solution:

$$\left(\frac{9}{5}\right)C=\left(\frac{9}{5}\right)\frac{5}{9}(F-32)$$ Multiplicative inverse; $\frac{9}{5}$ is the reciprocal of $\frac{5}{9}$

$$\frac{9}{5}C \quad = F-32$$ Simplify

$$+32 \quad +32$$ Additive inverse; add 32 to both sides

$$\frac{9}{5}C+32=F$$ Simplify

Different Types of Solutions for Linear Equations

In the last few examples, we solved for the *one* numerical value that makes the mathematical statement true. Some equations, however, will never be true no matter what value is assigned to the variable, and may have **no solution.** Other equations may be true for any value that is assigned to the variable and, therefore, have **infinitely many** solutions.

Let's try another type of example.

Example 14:

Solve $3x + 5 + 6x = 4 + 9x - 1$.

Solution:

$$3x+5+6x=4+9x-1$$

$$9x+5=3+9x$$ Combine like terms

$$-9x \quad -9x$$ Additive inverse; collect variables on one side

$$5=3$$ Simplify

Note that the simplified expressions on either side of the equal sign are NOT equivalent. This means that although the two expressions were written as an equation, they were never equal to begin with.

When this occurs, you can conclude that there is no value for x that will ever make this mathematical statement true. Therefore, there is no solution.

Example 15:

Solve $4 + 5x + 6 = 5(x + 2)$.

Solution:

$4+5x+6 = 5(x+2)$	
$5x+10=5x+10$	Combine like terms; distributive property
$-5x \quad -5x$	Collect variables on one side; subtract $5x$ from both sides
$10=10$	

In this case, note that after simplifying, the expressions on both sides of the equal sign are identical. In Example 14, no value of x could make this statement true. But here in Example 15, ANY value of x would make this statement true.

You can say that there are ***infinitely many solutions*** and x is ***all real numbers.***

When solving equations, there are three different possible types of solutions: *a finite number of solutions, no solution,* or *infinitely many solutions.*

Example 16:

How many solutions are there to this equation?

$$11x + 8 - 3x = 2(4x - 5)$$

A. One solution
B. No solution
C. Two solutions
D. Infinitely many solutions

Solution:

$$11x+8-3x=2(4x-5)$$

$8x+8=2(4x-5)$	Combine like terms on left side
$8x+8=8x-10$	Distributive property; multiply by 2
$-8x \quad -8x$	Collect variables on one side; additive inverse
$8= \quad -10$	
$8\neq-10$	Choice B, no solution

TECH TIPS: Solving Equations with a Graphing Calculator

Recall that an equation is a mathematical statement that two expressions are ***equal*** to each other. Any equation can be viewed as two separate expressions on opposite sides of the equal sign.

$$2x-4=-3x+6$$

Expression on left	**Expression on right**
$2x-4$	$-3x+6$

There is some value of x that will make the expression on the left side of the equal sign be equal to the expression on the right side of the equal sign.

Solving using a table

To find that value of x, let's compare the graphs and tables of both expressions.

Example:

Solve for x: $2x-4=-3x+6$

Solution:

1. Press **Y=** and enter the expression on the left into Y_1=.

   Enter the expression on the right into Y_2=.

Plot 1 Plot 2 Plot 3
\\Y_1= $2x - 4$
\\Y_2= $-3x + 6$
\\Y_3=
\\Y_4=
\\Y_5=
\\Y_6=
\\Y_7=

2. Press **2nd** **GRAPH** to view the table.

   The *y*-values for the expression on the *left* are listed under Y_1.

   The *y*-values for the expression on the *right* are listed under Y_2.

3. Look for the value of *x* that will make both expressions equal.

   Notice that when $x = 2$, the value of both expressions is the same.

X	Y_1	Y_2
0	−4	6
1	−2	3
2	0	0
3	2	−3
4	4	−6
5	6	−9
6	8	−12

X=2

4. Press **GRAPH** or **ZOOM6** to set your window to the standard parameters of –10 to 10 for *x* and *y*.

   Notice that the graphs intersect at a point.

   Press **TRACE**.

   All points on the line will now be shown at the bottom of the screen as you press the left and right arrow keys and trace along the line.

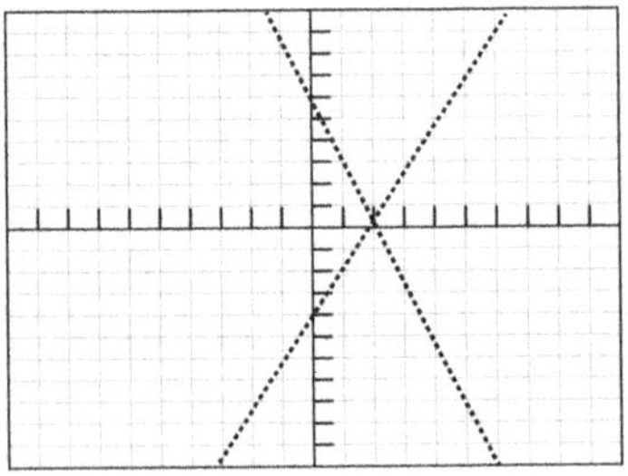

X = 3.8709677 Y = 3.7419355

5. While still in trace mode, press [2] [ENTER].

   The screen will display the value of the function at $x = 2$.

   Notice that the blinking cursor is highlighting the point where the two lines intersect.

   This confirms that both expressions are equal at $x = 2$.

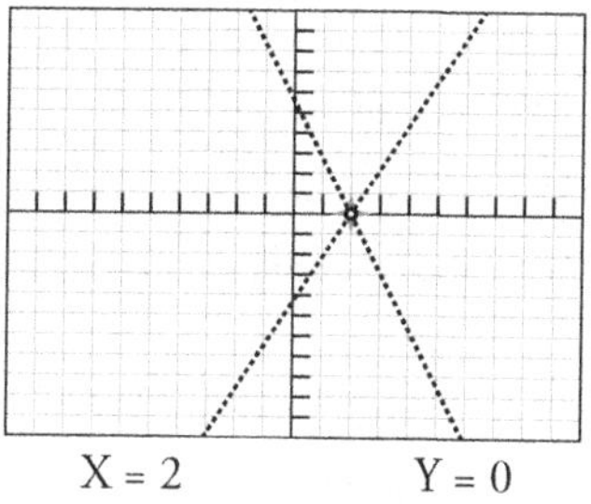

Answer: $x = 2$

Solving using a graph

Sometimes it's helpful to use the graph to solve for a variable.

Example:

Solve for x: $4x + 5 = -x + 1$.

Solution:

1. Press [Y=] and enter the expression on the left into Y_1=.

   Enter the expression on the right into Y_2=.

Plot 1 Plot 2 Plot 3
\Y_1= $4x + 5$
\Y_2= $-x + 1$
\Y_3=
\Y_4=
\Y_5=
\Y_6=
\Y_7=

2. Press [2nd] [GRAPH] to view the table.

   The y-values for the expression on the *left* are listed under Y_1.

   The y-values for the expression on the *right* are listed under Y_2.

3. Look for the value of x that will make both expressions equal.

X	Y_1	Y_2
−3	−7	4
−2	−3	3
−1	1	2
0	5	1
1	9	0
2	13	−1
3	17	−2

X=−3

4. Press the up arrow key.

   At $x = -1$, the value of Y_2 is greater than the value of Y_1.

   At $x = 0$, the value of Y_1 is greater than the value of Y_2.

   The value of x where both expressions are equal must be between $x = 1$ and $x = 2$.

   When the solution is not visible on the table, it is helpful to use the graph.

5. Press **GRAPH**.

   Notice that the graphs intersect at a point in Quadrant II.

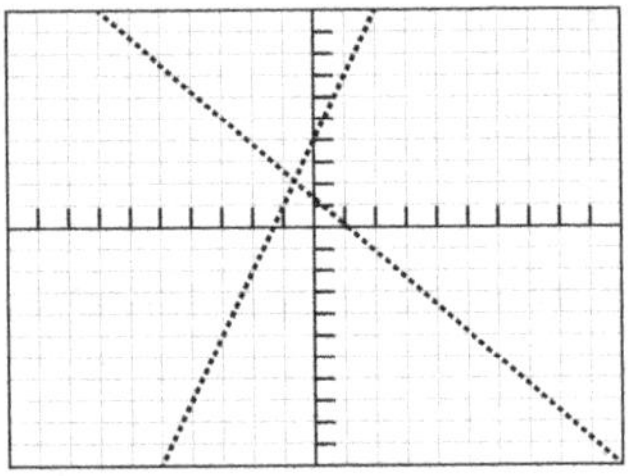

6. Press **2nd** **TRACE**.

7. While in CALC mode, press **5** to calculate the point of intersection.

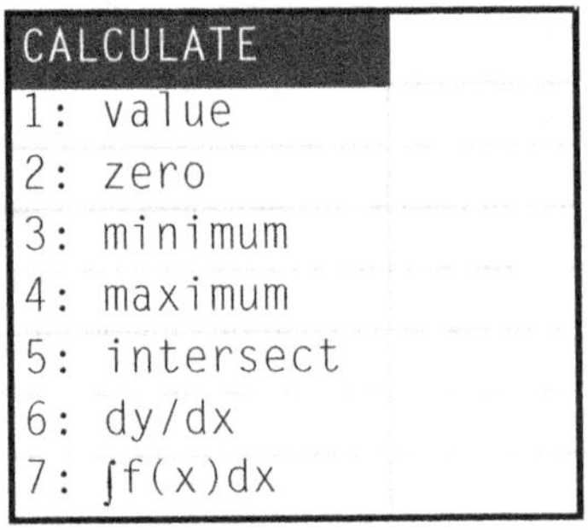

8. The screen will prompt you with three questions:

   First curve?
   Second curve?
   Guess?

   Press **ENTER** three times.

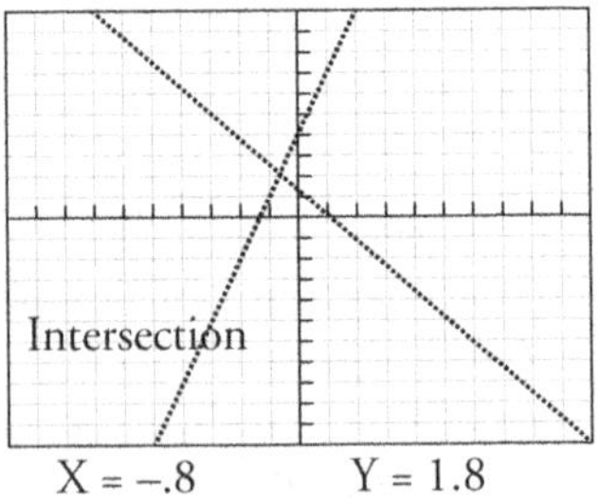

9. The point of intersection will appear at the bottom of the screen.

Answer: $x = -0.8$

Chapter Check-Out

Questions

For questions 1–9, solve for the variable in each equation.

1. $4 - m = 16$
2. $-7 + p = 21$
3. $l - \frac{5}{6} = \frac{2}{3}$
4. $-(y - 13) = 2y + 22$
5. $8g = 42$
6. $-\frac{b}{13} = 2.6$
7. $\frac{4}{7}h = 8$
8. $9x - 2 = 16$
9. $\frac{k}{15} - 4 = 1$
10. Jacob wants to buy a new pair of soccer cleats on sale for $74.99 after tax. So far, he has saved up $38 babysitting his little sister. Write and solve an equation to find how much money he still needs, *m*, to purchase the shoes.
11. Vertical angles are congruent. Find the value of *a* in the marked congruent angles below.

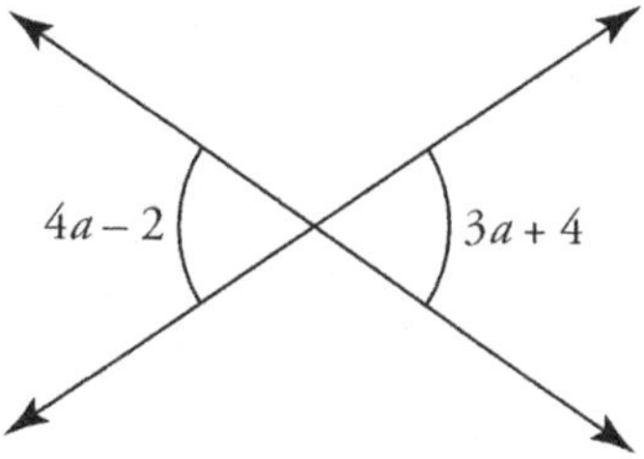

For questions 12–13, solve for the indicated variable.

12. Solve for *w*: $P = 2l + 2w$.
13. Solve for *a*: $F = ma$.

14. Flo's Pizza Factory charges \$8 for a plain cheese pizza and \$.50 per additional topping. Primo's Pizza Pies charges \$6 for a plain cheese pizza and \$1 per additional topping. Luis paid the same cost while ordering from both restaurants. How many toppings did Luis add to each pizza order?

For questions 15–16, determine if each equation has one solution, no solution, or infinitely many solutions.

15. $14z + 49 - 3z = 11(z + 5)$

16. $18 - (v + 2) = 4v + 16 - 5v$

Answers

1. $m = -12$
2. $p = 28$
3. $l = 1.5$
4. $y = -3$
5. $g = 5.25$
6. $b = -33.8$
7. $h = 14$
8. $x = 2$
9. $k = 75$
10. $38 + m = 74.99$; $m = \$36.99$
11. $a = 6$
12. $w = \frac{P}{2} - l$
13. $a = \frac{F}{m}$
14. 4
15. no solution
16. infinitely many solutions

Chapter 4
INEQUALITIES

Chapter Check-In

- ❑ Inequalities and their graphs
- ❑ Solving one-step inequalities
- ❑ Solving two-step and multi-step inequalities
- ❑ Inequalities with variables on both sides
- ❑ Compound inequalities
- ❑ Domain and range inequalities

Recall that an **equation** is a mathematical statement that two expressions are equal.

$$a = b$$

a is equal to b

An **inequality** is a statement that two expressions are not equal.

$$a \neq b$$

a is not equal to b

When comparing two values or expressions that are not equal, the following inequality symbols can also be used:

$a < b$	$a \leq b$
a is less than b	a is less than or equal to b
$a > b$	$a \geq b$
a is greater than b	a is greater than or equal to b

Inequalities and Their Graphs

A **solution of an inequality** is any value that makes the inequality true.

Example 1:

Determine which of the following is a solution of $x \leq 2$. Indicate all that apply.

A. –4

B. 6

C. 2

Solution:

A. $-4 \leq 2$, true. –4 is less than 2 and is a solution.

B. $6 \leq 2$, false. 6 is not less than 2.

C. $2 \leq 2$, true. 2 is less than or equal to 2 and is a solution.

Both Choices A and C are solutions.

Example 2:

Determine which of the following values is a solution of $7 + 3x > -5$.

A. 1

B. –6

Solution:

A.

$$\begin{aligned} 7+3(1) &> -5 && \text{Substitute 1 for } x \\ 7+3 &> -5 && \text{Simplify} \\ 10 &> -5 && \text{True; 1 is a solution} \end{aligned}$$

B.

$$\begin{aligned} 7+3(-6) &> -5 && \text{Substitute} -6 \text{ for } x \\ 7-18 &> -5 && \text{Simplify} \\ -11 &> -5 && \text{False;} -6 \text{ is not a solution} \end{aligned}$$

Only Choice A is a solution.

Graphing solutions on a number line

You can graph the solutions to an inequality on a number line.

Example 3:

Graph the following inequalities.

A. $x \geq -2$

B. $p < 4$

C. $1 \leq t$

Solution:

A. Closed dot on –2, shade right.

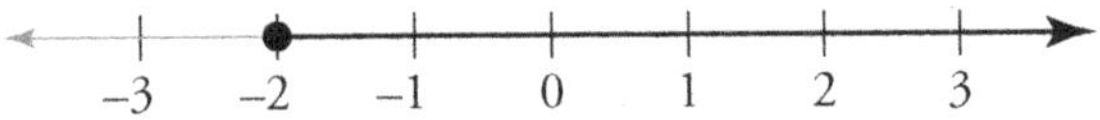

The closed dot indicates that –2 is a solution, and so is every value to the right.

B. Open dot on 4, shade left.

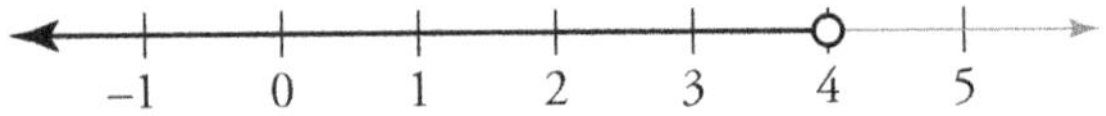

The open dot indicates that 4 is *not* a solution, but every value to the left is.

C. Can also be written as $t \geq 1$. Closed dot on 1, shade right.

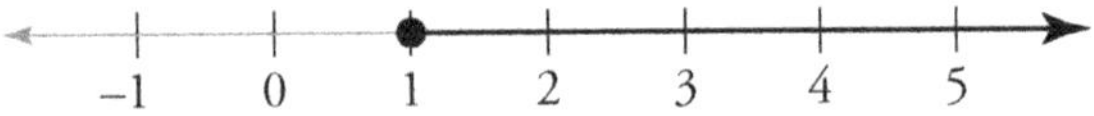

The closed dot indicates that 1 is a solution, and so is everything to the right.

Example 4:

Write an inequality for each situation.

A.

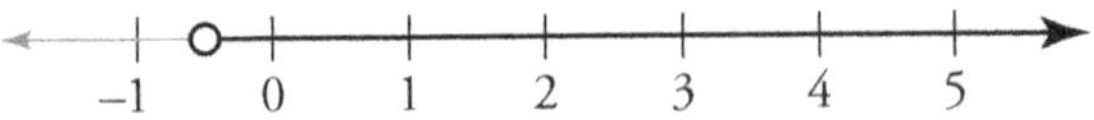

B.

C. The thermostat at Canyon Hills Middle School is set to 74°F in the warmer months. Identify a variable and write an inequality for a temperature that would cause the air conditioning system to turn on.

Solution:

A. The open dot indicates that –0.5 is *not* included. Shading to the right indicates that all values *greater than* –0.5 are solutions.

$x > -0.5$

B. The closed dot indicates that 2 is included. Shading to the left indicates that all values *less than* 2 are also solutions.

$x \le 2$

C. Let t = temperature in degrees Fahrenheit.

Temperatures *greater than and including* 74° would activate the air conditioning system.

$t \ge 74$

Solving One-Step Inequalities

To solve an inequality, much like an equation, isolate the variable using inverse operations and the properties of inequality.

Using addition and subtraction

Addition and Subtraction Properties of Inequality:

Addition	Subtraction
If $a < b$, then $a + c < b + c$.	If $a < b$, then $a - c < b - c$.

$$\begin{aligned} 4 &< 5 \quad \text{True} \\ 4+6 &< 5+6 \\ 10 &< 11 \quad \text{Still true} \end{aligned}$$

$$\begin{aligned} 7 &< 12 \quad \text{True} \\ 7-4 &< 12-4 \\ 3 &< 8 \quad \text{Still true} \end{aligned}$$

Note: These addition and subtraction properties are true for all inequality symbols: <, >, ≤, and ≥.

Example 5:

Solve each inequality and graph the solution.

A. $b - 5 > 8$

B. $g + 11 \leq 4$

C. Kilie gets a \$25 gift card to download apps and music on her smart phone. She has already spent \$18 downloading her favorite album. Write and solve an equation showing the possible price of apps or music files she can still purchase with the credit remaining on her gift card.

Solution:

A.

$$\begin{array}{rl} b-5>8 & \\ +5\ \ +5 & \text{Add 5 to both sides} \\ \hline b>13 & \text{Simplify} \end{array}$$

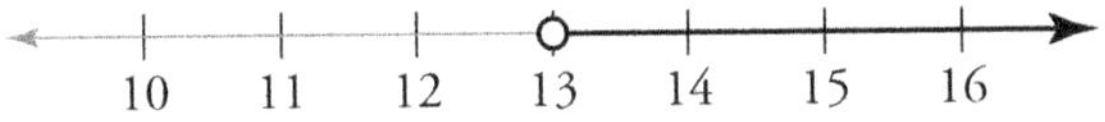

B.

$$\begin{array}{rl} g+11\leq 4 & \\ -11\ \ -11 & \text{Subtract 11 from both sides} \\ \hline g\leq -7 & \text{Simplify} \end{array}$$

C.

$$\begin{array}{rl} 18+c\leq 25 & \\ -18\qquad -18 & \text{Subtract 18 from both sides} \\ \hline c\leq 7 & \text{Simplify} \end{array}$$

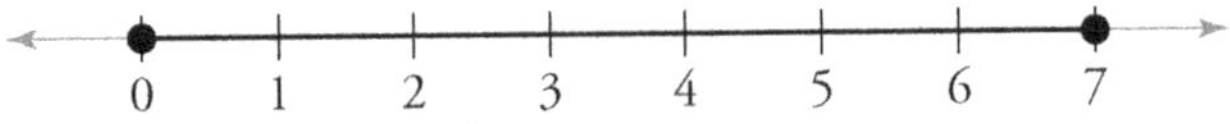

Kilie can purchase apps or music files priced at 7 *dollars or less*. Notice that the shading to the left stops at 0 because she cannot spend negative amounts of money.

Using multiplication and division

Multiplication and Division Properties of Inequality for $c > 0$:

Multiplication	Division
If $a < b$, then $ac < bc$.	If $a < b$, then $\frac{a}{c} < \frac{b}{c}$.

$2 < 3$ True	$6 < 8$ True
$2(7) < 3(7)$	$\frac{6}{2} < \frac{8}{2}$
$14 < 21$ Still true	$3 < 4$ Still true

Note: These multiplication and division properties are true for all inequality symbols: <, >, ≤, and ≥.

When multiplying or dividing by a negative number, notice that the inequality is no longer true.

$6 < 10$ True	$9 < 18$ True
$6(-1) < 10(-1)$	$\frac{9}{-3} < \frac{18}{-3}$
$-6 < -10$ No longer true	$-3 < -6$ No longer true

In order to make the inequality true, you must reverse the inequality symbol.

$-6 > -10$ True $\qquad -3 > -6$ True

Multiplication and Division Properties of Inequality for $c < 0$:

Multiplication	Division
If $a < b$, then $a(c) > b(c)$.	If $a < b$, then $\frac{a}{c} > \frac{b}{c}$.

Note: These multiplication and division properties are true for all inequality symbols: <, >, ≤, and ≥.

Example 6:

Solve each inequality and graph the solution.

A. $\frac{m}{7} \geq 2$

B. $\frac{2}{3}k < 6$

C. $-\frac{y}{6} \leq 1$

D. $21 > -3p$

E. Myrna's professional tutoring business charges $40 an hour for test-prep sessions. Myrna wants to purchase a 3-day pass for the Austin City Limits music festival for $240. Write and solve an inequality showing the fewest number of clients, *c*, she needs to tutor to be able to afford her 3-day festival pass.

Solution:

A.

$$\frac{m}{7} \geq 2$$

$$(7)\frac{m}{7} \geq 2(7) \quad \text{Multiply both sides by 7}$$

$$m \geq 14$$

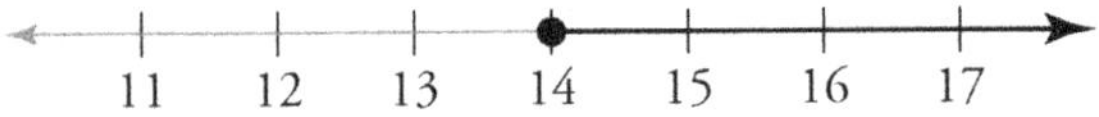

B.

$$\frac{2}{3}k < 6$$

$$\left(\frac{3}{2}\right)\frac{2}{3}k < \frac{6}{1}\left(\frac{3}{2}\right) \quad \text{Multiply by } \frac{3}{2} \text{ on both sides}$$

$$k < \frac{18}{2} \quad \text{Simplify}$$

$$k < 9$$

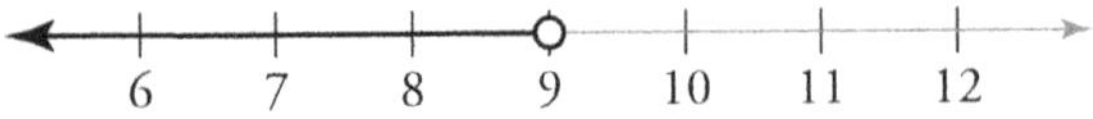

C. $-\frac{y}{6} \le 1$

$(-6)\left(-\frac{y}{6}\right) \ge 1(-6)$ Multiply both sides by – 6; reverse inequality symbol

$y \ge -6$ Simplify

D. $21 > -3p$

$\frac{21}{-3} < \frac{-3p}{-3}$ Divide both sides by – 3; reverse inequality symbol

$-7 < p$ Simplify

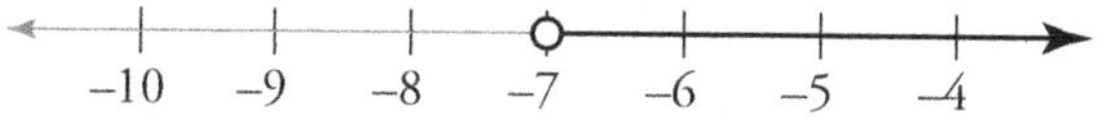

E. $40c \ge 240$

$\frac{40c}{40} \ge \frac{240}{40}$ Divide both sides by 40

$c \ge 6$ Simplify

Myrna must tutor at least 6 clients to pay for her pass.

Solving Two-Step and Multi-Step Inequalities

When you evaluate an expression with more than one operation, you need to follow the order of operations. However, since you are trying to "undo" these operations to solve for a variable, you must work backward. Undo the addition and subtraction first, and then undo the multiplication and division.

Example 7:

Solve and graph the solution to the inequality $18 + 9z \le 225$.

Solution:

$$\begin{aligned} 18+9z &\leq 225 & \\ -18 \qquad -18 & & \text{Subtract 18 from both sides} \\ 9z &\leq 207 & \text{Simplify} \\ \frac{9z}{9} &\leq \frac{207}{9} & \text{Divide both sides by 9} \\ z &\leq 23 & \text{Simplify} \end{aligned}$$

Some inequalities need to be simplified first before solving. Often, you may be able to use the distributive property and/or combine like terms before beginning to isolate the variable.

Example 8:

Solve and graph the solution to the inequality $8(x-3)+5x>-167$.

Solution:

$$\begin{aligned} 8(x-3)+5x &> -167 & \\ 8x-24+5x &> -167 & \text{Distributive property} \\ 13x-24 &> -167 & \text{Combine like terms} \\ +24 \qquad +24 & & \text{Add 24 to both sides} \\ 13x &> -143 & \text{Simplify} \\ \frac{13x}{13} &> \frac{-143}{13} & \text{Divide both sides by 13} \\ x &> -11 & \text{Simplify} \end{aligned}$$

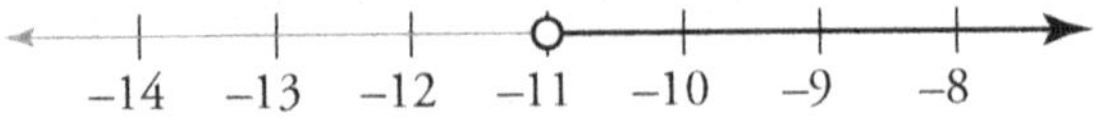

Inequalities with Variables on Both Sides

To solve an inequality with variables on both sides, you must first move all the variables to one side of the inequality.

Example 9:

Solve and graph the solution to the inequality $3r + 11 < -5r + 3$.

Solution:

$3r + 11 < -5r + 3$	
$+5r \qquad +5r$	Move variables to one side; add $5r$ to both sides
$8r + 11 < 3$	Simplify
$-11 \quad -11$	Subtract 11 from both sides
$8r < -8$	Simplify
$\frac{8r}{8} < \frac{-8}{8}$	Divide by 8 on both sides
$r < -1$	Simplify

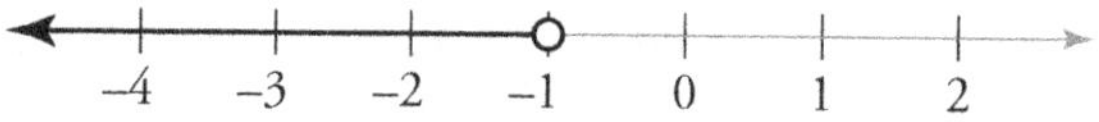

Example 10:

Caleb and Melissa are planning to rent a car for a road trip to visit their grandmother in El Paso, Texas. Caleb wants to use Company A that charges an initial fee of $32 and an additional $3 per day. Melissa wants to use Company B that charges an initial fee of $48 and an additional $1 per day. Write and solve an inequality showing the number of days, *d*, they'll need to rent a car in order for Company A to be less expensive than Company B.

Solution:

$$
\begin{array}{ll}
32+3d<48+d & \\
\quad -d < \quad -d & \text{Subtract } d \text{ from both sides} \\
32+2d<48 & \text{Simplify} \\
-32 \qquad < -32 & \text{Subtract 32 from both sides} \\
2d<16 & \text{Simplify} \\
\frac{2d}{2}<\frac{16}{2} & \text{Divide both sides by 2} \\
d<8 & \text{Simplify}
\end{array}
$$

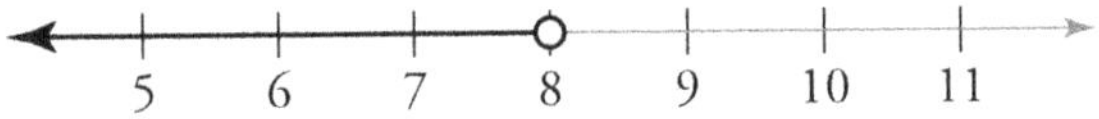

Company A will be less expensive than Company B if they decide to rent a car for less than 8 days.

Compound Inequalities

In Example 10, it wouldn't make sense for *all values less than 8* to be possible solutions since that would also include all negative numbers. In a real-world situation, it wouldn't be reasonable to rent a car a *negative* number of days.

A more reasonable solution would be to say *all values less than 8* **and** *greater than 0, or* $d > 0$ *and* $d < 8$.

Two inequalities can be combined into one statement as a **compound inequality** using the word *AND* or the word *OR*.

$$0 < d < 8$$

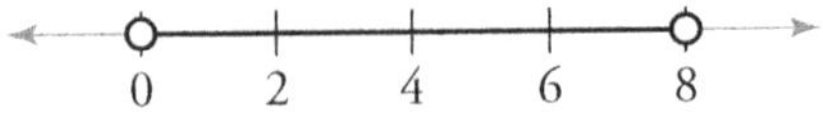

Graphs of compound inequalities with *AND*

Example 11:

Write the following as compound inequalities and graph the solution.

A. $x \geq -1$ *AND* $x < 3$

B. $x > -4$ *AND* $x < 2$

Solution:

A. $-1 \leq x < 3$

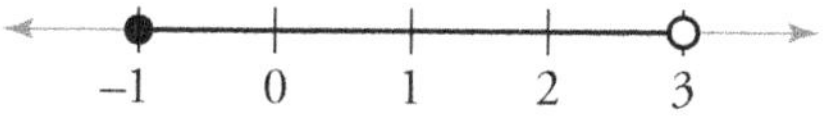

B. $-4 < x < 2$

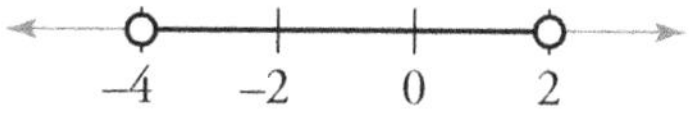

Notice that the graphs of compound inequalities using the word *AND* are shaded *between* the two numbers where the overlap of both inequalities occurs.

Graphs of compound inequalities with *OR*

Example 12:

Graph the solutions to the compound inequality using the word *OR*.

A. $x < 3$ *OR* $x \geq 5$

B. $x \leq 4$ *OR* $x > 7$

Solution:

A.

B.

Notice that the graphs of compound inequalities using the word *OR* are shaded in opposite directions.

Solving compound inequalities

To solve a compound inequality using *AND*, write as two separate inequalities and solve each separately.

Example 13:

Solve and graph $-3 \leq 7 + n < 8$.

Solution:

$$-3 \leq 7 + n < 8$$

$-3 \leq 7 + n$	*AND*	$7 + n < 8$	Write as two inequalities
$-7 \quad -7$		$-7 \quad -7$	Solve each inequality
$-10 \leq n$	*AND*	$n < 1$	
	$-10 \leq n < 1$		Write as a compound inequality

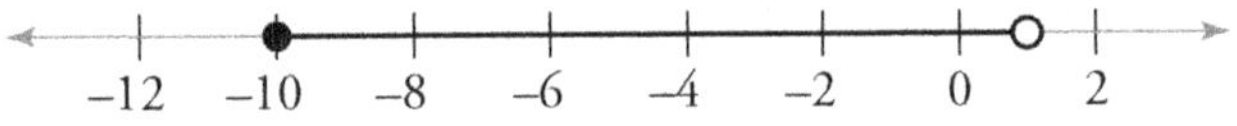

Example 14:

Solve and graph $\frac{g}{2} > 3$ *OR* $4g + 1 \leq 9$.

Solution:

$\frac{g}{2} > 3$	*OR*	$4g + 1 \leq 9$	
$(2)\frac{g}{2} > 3(2)$		$-1 \quad -1$	Solve each inequality.
$g > 6$		$4g \leq 8$	
		$\frac{4g}{4} \leq \frac{8}{4}$	
		$g \leq 2$	
$g > 6$	*OR*	$g \leq 2$	

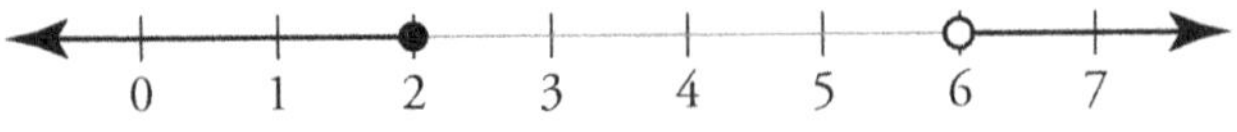

Domain and Range Inequalities

Domain and range can be represented using inequalities. Recall from Chapter 2 that **domain** is the set of all input values in a function and **range** is the set of all output values in a function.

Reasonable domain and range

In real-life situations, it would be considered reasonable to use certain types of numbers and not others. For example, consider the relationship between the number of hours you work, h, in a 40-hour workweek and your paycheck amount, $f(h)$, if your hourly wage is $9.

A **reasonable domain** would include positive real numbers since it is *unreasonable* to work a negative amount of hours. A **reasonable range** would include positive rational numbers since money can also be counted in decimals (cents).

	Verbal Description		***Inequality***
Domain	All real numbers between 0 and 40		$0 \leq h \leq 40$
Range	*Minimum* 9(0) = 0	*Maximum* 9(40) = 360	$0 \leq f(h) \leq 360$
	All rational numbers between 0 and 360		

Reasonable domain and range:

Domain: $0 \leq h \leq 40$

Range: $0 \leq f(h) \leq 360$

Example 15:

Taylor's new car gets an estimated 28 miles to the gallon. A full gas tank holds 13 gallons of gasoline. What are all the reasonable numbers of miles, m, that Taylor's car could run on g gallons of gasoline before having to refuel?

Solution:

Domain	All real numbers between 0 and 13		$0 \leq g \leq 13$
Range	*Minimum* 28(0) = 0	*Maximum* 28(13) = 364	$0 \leq m \leq 364$

Domain: $0 \leq g \leq 13$

Range: $0 \leq m \leq 364$

Domain and range for a graph

Inequalities can also be used to describe the domain and range for a graph.

Example 16:

Identify the domain and range for each graph.

A.

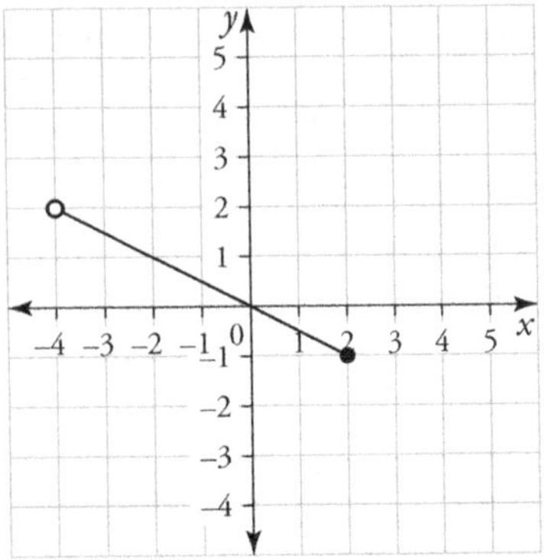

B.

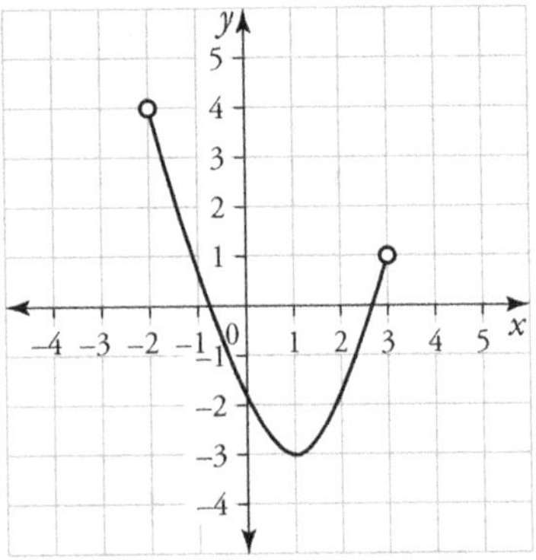

Solution:

A. From left to right on the x-axis, the graph goes from –4 to 2.
–4 is not included, open dot
2 is included, closed dot
Domain: $-4 < x \leq 2$

From bottom to top on the y-axis, the graph goes from –1 to 2.
–1 is included, closed dot
2 is not included, open dot
Range: $-1 \leq y < 2$

B. From left to right on the x-axis, the graph goes from –2 to 3.
–2 is not included, open dot
3 is not included, open dot
Domain: $-2 < x < 3$

From bottom to top on the y-axis, the graph goes from –3 to 4.
–3 is included
4 is not included, open dot
Range: $-3 \leq y < 4$

Some graphs continue infinitely in one or more directions when denoted by an arrow.

Example 17:

Identify the domain and range for each graph.

A.

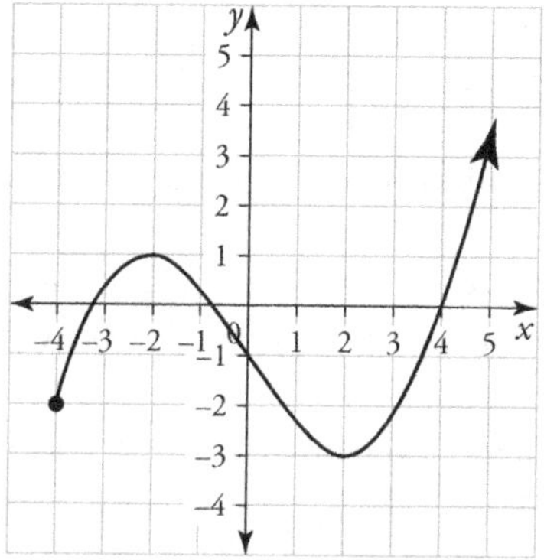

B.

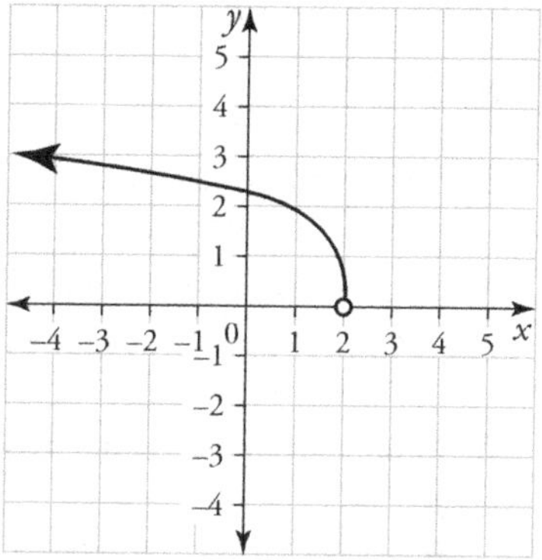

Solution:

A. From left to right on the *x*-axis, the graph goes from –4 and includes all numbers greater than –4.
–4 is included, closed dot
arrow to the right, approaches positive infinity
Domain: $x \geq -4$

From bottom to top on the *y*-axis, the graph goes from –3 and includes all numbers greater than –3.
–3 is included
arrow up, approaches positive infinity
Range: $y \geq -3$

B. From left to right on the *x*-axis, the arrow shows the graph continues infinitely to the left, including all numbers less than 2.
2 is not included, open dot
Domain $x < 2$

From bottom to top on the *y*-axis, the graph goes from 0 and continues infinitely up including all values greater than 0.
0 is not included, open dot
Range: $y > 0$

Chapter Check-Out

Questions

For questions 1–4, write the inequality represented by each situation.

1. 

2. 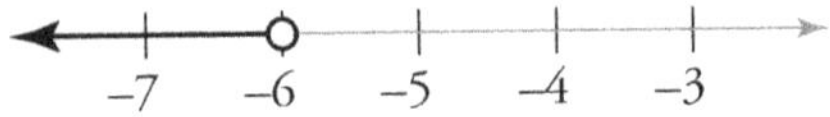

3. A high school elective class will only be offered if at least 24 students register for that class.

4. In order to lose weight while exercising, a nutrition expert recommends that her client keep her heart rate above 135 beats per minute.

5. Which value of x is in the solution set to the inequality $-3x + 19 < 4$?

A. 5
B. –2
C. 7
D. –5

For questions 6–8, solve each inequality using addition or subtraction. Graph the solution.

6. $w - 4 > 3.2$

7. $6 \geq k + 8$

8. $\frac{2}{3} + h \leq \frac{1}{2}$

For questions 9–11, solve each inequality using multiplication or division. Graph the solution.

9. $-12 < 4b$

10. $\frac{a}{9} > -11$

11. $\frac{3}{7}v \geq 6$

For questions 12–17, solve each inequality and graph the solution.

12. $8(-2x + 15) < 24$

13. $\frac{1}{3}r + \frac{1}{2} \geq \frac{2}{3}$

14. $\frac{l+6}{5} > 3$

15. $6 + 2j \geq -4j$

16. $3(y - 4) > -5(y + 8)$

17. Rod opens a savings account with a \$180 deposit and will continue to deposit \$15 per month. Carolina's savings account has a balance of \$125, and she will deposit \$20 per month. Write and solve an inequality that will show how many months it will take for Carolina's account balance to be greater than Rod's account balance.

For questions 18–20, solve each compound inequality and graph the solution.

18. $-7 < \frac{c}{2} + 4 \leq 9$

19. $13 < 6f + 1$ *AND* $4f - 19 \leq 5$

20. $\frac{b}{9} < 0$ *OR* $-3b + 8 < 2$

Answers

1. $x \geq 6$
2. $x < -6$
3. $s \geq 24$
4. $b > 135$
5. C
6. $w > 7.2$

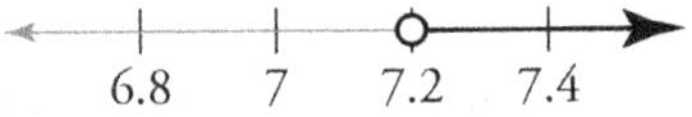

7. $k \leq -2$

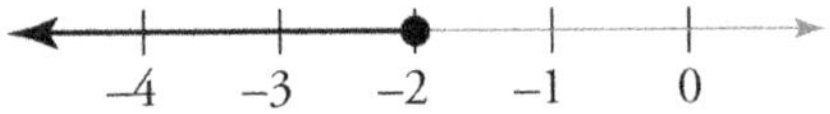

8. $h \leq -\frac{1}{6}$

9. $b > -3$

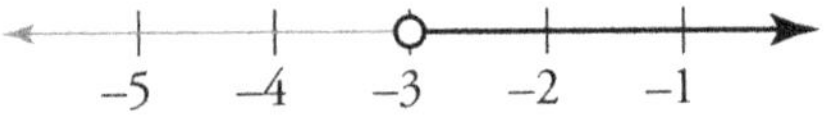

10. $a > -99$

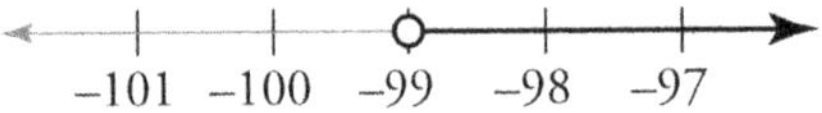

11. $v \geq 14$

12. $x > 6$

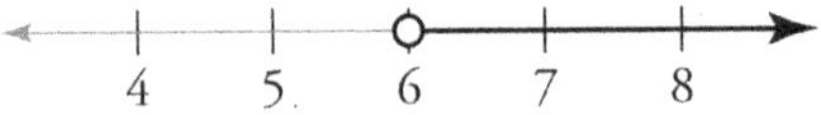

13. $r \geq \frac{1}{2}$

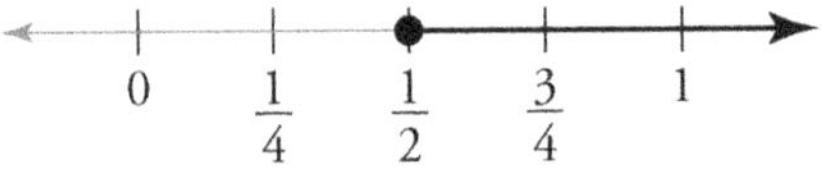

14. $l > 9$

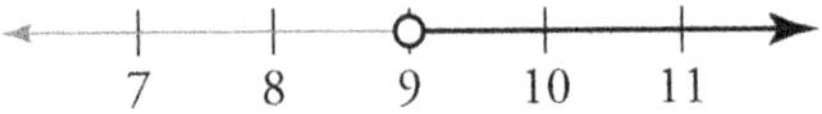

15. $j \geq -1$

16. $y > -3.5$

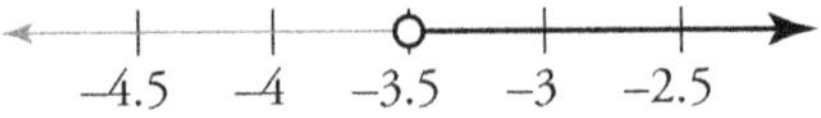

17. $180 + 15x < 125 + 20x$; $x > 11$ months

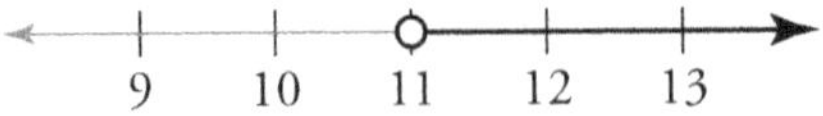

18. $-22 < c \leq 10$

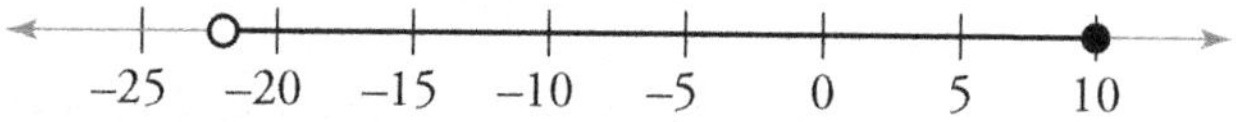

19. $2 < f \leq 6$

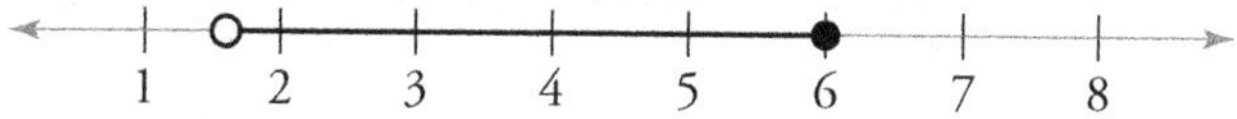

20. $b < 0$ *OR* $b > 2$

Chapter 5

LINEAR FUNCTIONS

Chapter Check-In

- ❑ Rate of change
- ❑ Slope
- ❑ Direct variation
- ❑ Slope-intercept form
- ❑ Point-slope form
- ❑ Parallel lines
- ❑ Perpendicular lines
- ❑ Standard form
- ❑ Linear parent function
- ❑ Scatter plots
- ❑ TECH TIPS: Creating scatter plots on a graphing calculator

Functions can be used to represent many real-world situations. In the remaining chapters, we will explore the different types of functions you can encounter in algebra I. A situation that involves a constant rate of change can be described by a **linear function.**

Rate of Change

Rate of change is a ratio that shows how a dependent variable changes in relation to how an independent variable changes.

$$\text{rate of change} = \frac{\text{change in the dependent variable}}{\text{change in the independent variable}}$$

Rate of change can be seen in many real-life situations.

Finding rate of change from a table

Example 1:

Consider the table below that shows the charges for Wi-Fi service after purchasing a new router. What is the rate of change? What does the rate of change represent?

Customer Bill	
Number of Months	***Charge ($)***
1	$82
2	$114
3	$146
4	$178

Solution:

Step 1. Identify the dependent and independent variables. The fee depends on the number of months you use the Wi-Fi service.

$$\text{rate of change} = \frac{\text{change in cost}}{\text{change in number of months}}$$

Step 2. Calculate the change in both variables by subtracting data points.

$$\text{From month 1 to month 2} = \frac{\text{change in cost}}{\text{change in number of months}}$$

$$= \frac{114-82}{2-1} = \frac{32}{1}$$

$$\text{From month 2 to month 3} = \frac{\text{change in cost}}{\text{change in number of months}}$$

$$= \frac{146-114}{3-2} = \frac{32}{1}$$

$$\text{From month 3 to month 4} = \frac{\text{change in cost}}{\text{change in number of months}}$$
$$= \frac{178 - 146}{4 - 3} = \frac{32}{1}$$

The rate of change is $\frac{32}{1}$. Notice that the rate of change is *the same* for each pair of consecutive data points. This means that the cost of Wi-Fi service increases by $32 for each additional month.

Example 2:

The table below shows the total distance a hummingbird flies when measured at different time intervals. Find the rate of change. What does the rate of change represent?

Time (seconds)	2	5	6	8
Distance (meters)	30	75	90	120

Solution:

$$\frac{75-30}{5-2} = \frac{45}{3} = \frac{15}{1} \qquad \frac{90-75}{6-5} = \frac{15}{1} \qquad \frac{120-90}{8-6} = \frac{30}{2} = \frac{15}{1}$$

The rate of change is $\frac{15}{1}$. The hummingbird is flying at a rate of 15 meters per second.

When the rate of change between all data points is the same, we say there is a **constant rate of change.**

Because the rate of change between points is constant, the graph forms a straight line. A function whose graph is a straight line is called a **linear function.**

Graphs of linear and non-linear relations

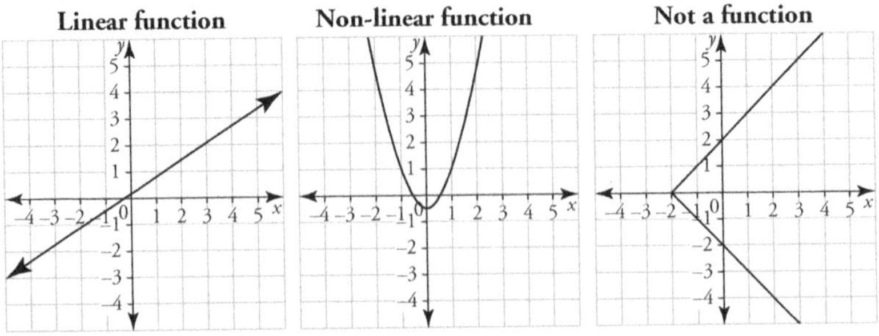

Example 3:

Identify the rate of change. Determine if the following data represents a linear function.

x	y
1	–18
3	–32
4	–39
7	–60

Solution:

Step 1. Calculate the difference between consecutive points on the table.

Change in x	x	y	Change in y
	1	–18	
2			–14
	3	–32	
1			–7
	4	–39	
3			–21
	7	–60	

If y is the dependent variable and x is the independent variable, then the rate of change can be expressed as follows:

$$\text{rate of change} = \frac{\text{change in the dependent variable}}{\text{change in the independent variable}} = \frac{\text{change in } y}{\text{change in } x}$$

$$\frac{-14}{2} = -7 \qquad \frac{-7}{1} = -7 \qquad \frac{-21}{3} = -7$$

Step 2. Compare the rates of change to determine if they are constant.

Rate of change = –7.

Since it is a constant rate of change, the function is linear.

Slope

The **slope** of a graph describes how steep it is between two points and whether it's rising or falling.

The steepness of a line can be characterized by how much it changes vertically compared to how it changes horizontally. The **rise** of a line describes its vertical change, or its change in the *y*-values of any two points on the graph. The **run** of a line describes its horizontal change, or its change in the *x*-values of any two points on the graph.

$$\text{slope} = \frac{\text{vertical change}}{\text{horizontal change}} = \frac{\text{rise}}{\text{run}}$$

Types of slope

There are four types of slope: positive, negative, zero, and undefined.

Positive slope

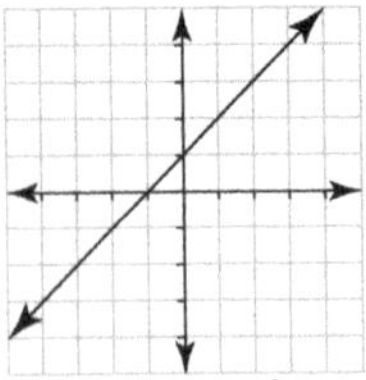

Lines that rise from left to right

Negative slope

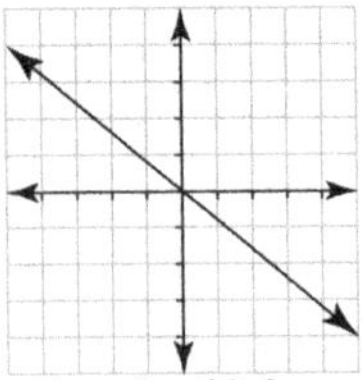

Lines that fall from left to right

Zero slope

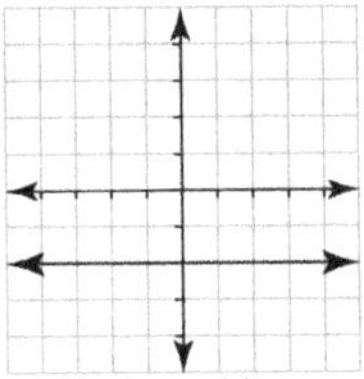

Horizontal lines

Undefined slope

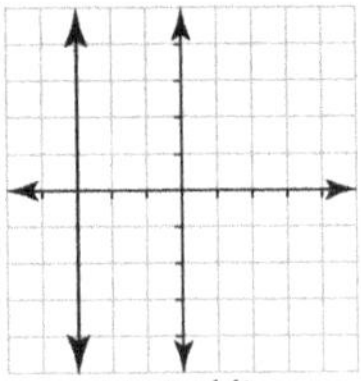

Vertical lines

Slopes of horizontal and vertical lines

Example 4:

Find the slope of each line.

A.

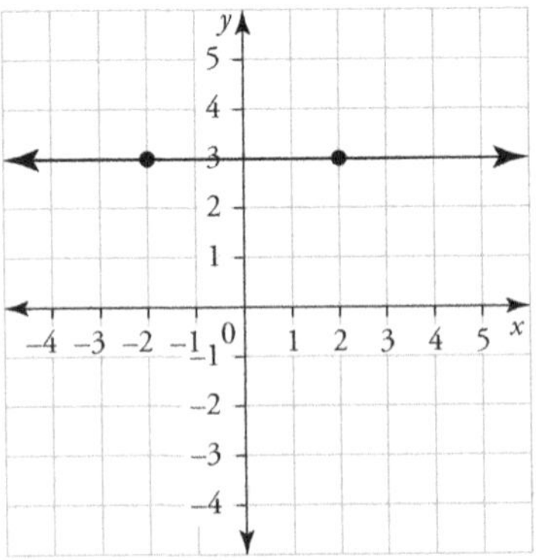

B.

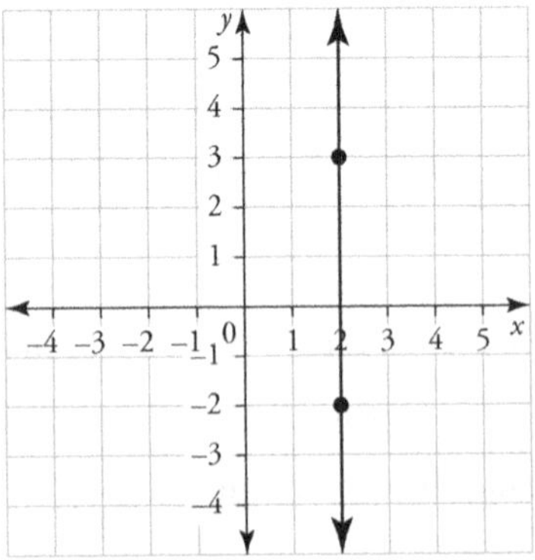

Solution:

A. $\text{slope} = \dfrac{\text{rise}}{\text{run}} = \dfrac{0}{4} = 0$

B. $\text{slope} = \dfrac{\text{rise}}{\text{run}} = \dfrac{5}{0}$. Division by 0 is undefined; therefore, slope = undefined.

Slope formula

You can find the slope of a line by using the coordinates of any two points, (x_1, y_1) and (x_2, y_2). Slope is often represented by the variable m.

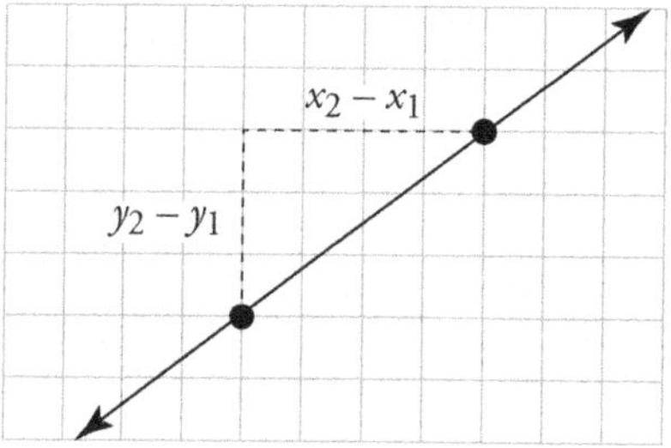

$$\text{slope} = m = \frac{\text{rise}}{\text{run}} = \frac{y_2 - y_1}{x_2 - x_1}$$

Example 5:

Find the slope of the line that contains the points (0, 3) and (8, 5).

Solution:

$m = \dfrac{y_2 - y_1}{x_2 - x_1}$

$m = \dfrac{5-3}{8-0}$ Substitute (0, 3) for (x_1, y_1) and (8, 5) for (x_2, y_2)

$m = \dfrac{2}{8}$ Simplify.

$m = \dfrac{1}{4}$

Finding slope from a graph

To find the rise, start at a point and count the vertical distance, from left to right, between any two points.

Next, to find the run, count the horizontal distance between the same two points, from left to right.

In the graph below, $\text{slope} = \frac{\text{rise}}{\text{run}} = \frac{2}{3}$.

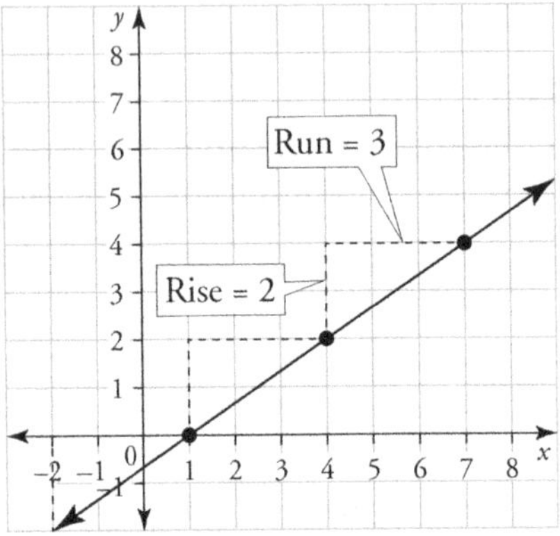

Measuring rise and run from a graph

Example 6:

Find the slope of each line.

A.

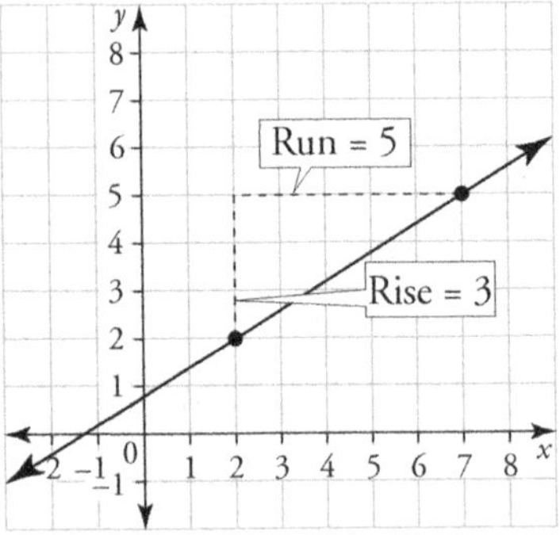

B.

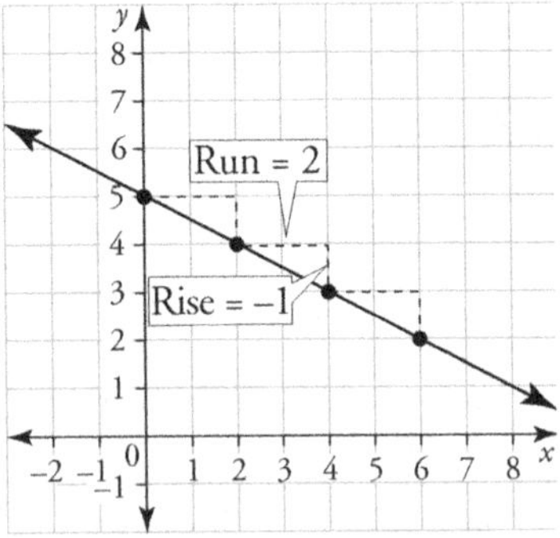

Solution:

A. $\text{slope} = \frac{\text{rise}}{\text{run}} = \frac{3}{5}$

B. $\text{slope} = \frac{\text{rise}}{\text{run}} = \frac{-1}{2}$

Note: The slope of the line that *rises* from left to right is *positive*, while the slope of the line that *falls* from left to right is *negative*.

Direct Variation

Grace decides to cook brisket for her son's graduation party. She estimates that 1 brisket will feed 8 of her guests. The following table and graph show how the number of guests served varies directly with the number of briskets.

Number of Briskets	***Number of Guests Served***
1	8
2	16
3	24
4	32
5	40

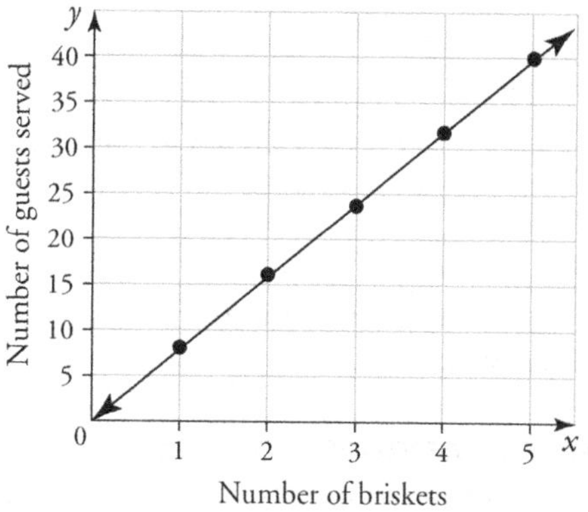

This relationship can be described by the equation $y = 8x$, where the number of guests Grace can serve *varies directly* with the number of briskets she cooks.

Direct variation is a linear relationship that can be written in the form $y = kx$, where k is the **constant of variation.**

Example 7:

For each of the following, determine if the equation is a direct variation. If it is, identify the constant of variation, k.

A. $y = -5x$

B. $5y = 3x$

C. $-4x + y = 7$

Solution:

A. The equation is written in the form $y = kx$, so the equation is a direct variation. The constant of variation is $k = -5$.

B. The equation can be written in the form $y = kx$, so the equation is a direct variation.

$$5y = 3x \quad \text{Solve the equation for } y$$

$$\frac{5y}{5} = \frac{3x}{5} \quad \text{Divide both sides by 5}$$

$$y = \frac{3}{5}x \quad \text{Simplify}$$

The constant of variation is $k = \frac{3}{5}$.

C.

$$-4x + y = 7 \quad \text{Solve the equation for } y$$

$$+4x \quad = +4x \quad \text{Add } 4x \text{ to both sides}$$

$$y = 4x + 7$$

The equation cannot be written in the form $y = kx$, so the equation is NOT a direct variation.

Writing a direct variation equation

You can rewrite a direct variation $y = kx$ as $k = \frac{y}{x}$ to solve for k, the constant of variation. For example,

$$y = kx$$

$$\frac{y}{x} = \frac{kx}{x} \quad \text{Divide both sides by } x$$

$$\frac{y}{x} = k$$

You can also use $k = \frac{y}{x}$ to determine if points on a table represent a direct variation by comparing each pair of input and output values. If each ratio $\frac{y}{x}$ is constant for every ordered pair, then it is a direct variation.

Example 8:

In the situations below, decide whether y varies directly with x. If it does, write the equation for the direct variation.

A. Find the ratio $\frac{y}{x}$ for each ordered pair.

x	1	3	6
y	4	12	24

B. Find the ratio $\frac{y}{x}$ for each ordered pair.

x	1	3	6
y	4	6	9

Solution:

A. $\frac{4}{1} = 4 \qquad \frac{12}{3} = 4 \qquad \frac{24}{6} = 4$

The ratio $\frac{y}{x}$ is constant for every ordered pair where $k = 4$; therefore, y varies directly with x. The equation for the direct variation is $y = 4x$.

B. $\frac{4}{1} = 4 \qquad \frac{6}{3} = 2 \qquad \frac{9}{6} = \frac{3}{2}$

This is not a direct variation because the ratio $\frac{y}{x}$ is NOT constant for every ordered pair.

Graphing a direct variation equation

Example 9:

A car travels 124 miles at a constant speed on 4 gallons of gas. Find the constant of variation k. Make a table of values showing the number of miles traveled, y, with each gallon of gasoline, x. Make a graph that represents this direct variation.

Solution:

The distance a car can travel depends on the amount of gasoline it contains. Distance is the dependent variable, and amount of gas is the independent variable.

Step 1. To find the constant of variation k, find the ratio $\frac{y}{x}$.

$$\frac{124}{4} = 31$$

This relationship can be written as

distance = 31 times number of gallons of gasoline

$$y = 31x$$

Step 2. Make a table of values and plot the points on a coordinate plane.

x	$y = 31x$	y
0	$y = 31(0)$	0
1	$y = 31(1)$	31
2	$y = 31(2)$	62
3	$y = 31(3)$	93
4	$y = 31(4)$	124

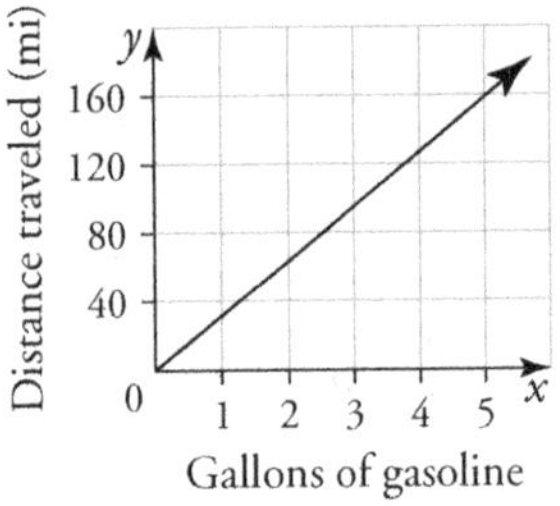

Example 10:

Graph the direct variation $y = 3x$.

Select input values for a table and evaluate. Plot on a coordinate grid.

Solution:

x	$y = 3x$	(x, y)
−2	3(−2) = −6	(−2, −6)
−1	3(−1) = −3	(−1, −3)
0	3(0) = 0	(0, 0)
1	3(1) = 3	(1, 3)
2	3(2) = 6	(2, 6)

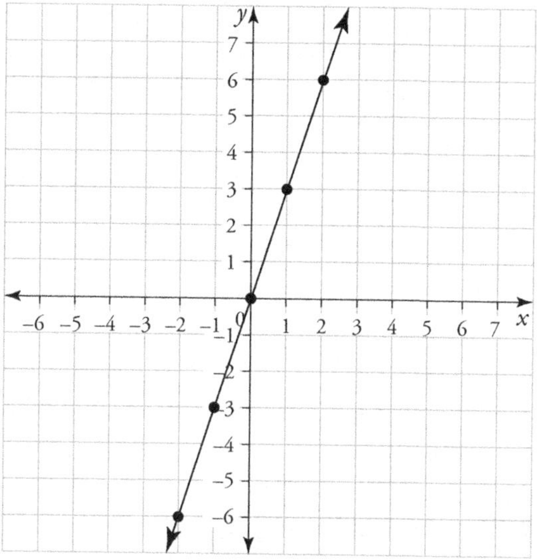

Notice that direct variation equations graph as straight lines. All direct variations are also linear functions.

Note: Graphs of direct variations always go through the origin, (0, 0).

Solving direct variation word problems

Example 11:

The amount of money your school's student council earns during a fundraiser dinner varies directly with the number of students who attend the fundraiser. If the student council raises $720 when 40 students attend the fundraiser, how much money will it raise if 75 students attend?

A. Identify the independent and dependent variables.

B. Find the constant of variation $\frac{y}{x}$.

C. Write an equation of variation.

D. Substitute 75 for x and solve for y.

Solution:

A. The amount of money raised depends on the number of students who attend. Money raised is the dependent variable, y, and number of students is the independent variable, x.

B. $\frac{720}{40} = 18$

$k = 18$

C. Money raised = 18 times the number of students in attendance

$$y = 18x$$

D. $y = 18(75)$

$y = \$1{,}350$

The student council will raise $1,350 if 75 students attend the fundraiser.

Slope-Intercept Form

The **slope-intercept form** of a linear equation is as follows, where m = slope and b = y-intercept.

$$y = mx + b$$

slope (m) y-intercept (b)

Graphs of linear functions

Compare the graphs of the following functions:

$y = x$		
x	$y = x$	(x, y)
−1	$y = -1$	(−1,−1)
0	$y = 0$	(0, 0)
1	$y = 1$	(1, 1)
2	$y = 2$	(2, 2)

$y = 4x$		
x	$y = 4x$	(x, y)
−1	$y = 4(-1)$	(−1, −4)
0	$y = 4(0)$	(0, 0)
1	$y = 4(1)$	(1, 4)
2	$y = 4(2)$	(2, 8)

$y = -2x + 3$		
x	$y = -2x + 3$	(x, y)
−1	$y = -2(-1) + 3$	(−1, 5)
0	$y = -2(0)+3$	(0, 3)
1	$y = -2(1)+3$	(1, 1)
2	$y = -2(2)+3$	(2, −1)

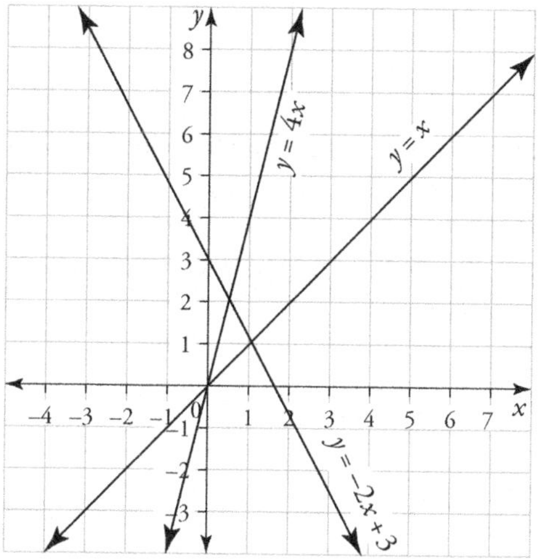

All three graphs represent linear functions.

The equation $y = x$ is the simplest form of a linear function. The simplest form of a function is called the **parent function.**

The function $y = 4x$ is a direct variation because it is written in the form $y = kx$ and because its graph passes through the origin, (0, 0).

The function $y = -2x + 3$ is not a direct variation because it is not written in the form $y = kx$ and because its graph does not *pass through the origin,* (0, 0).

The ***y*-intercept** is the point at which a line intersects the *y*-axis.

The function $y = -2x + 3$ has a *y*-intercept of 3 because it intersects the *y*-axis at (0, 3), and it has a slope of –2.

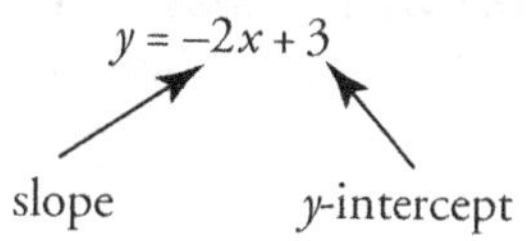

Identifying slope and *y*-intercept

Identify the slope and *y*-intercept of the following functions.

Example 12:

A. $y = \frac{2}{5}x - 7$

B. $2x + 3y = 12$

Solution:

A. $m = \frac{2}{5}$; $b = -7$

B. Isolate the variable y to rewrite the function in $y = mx + b$ form.

$$2x + 3y = 12$$

$$-2x \quad = -2x \qquad \text{Subtract } 2x \text{ from both sides}$$

$$3y = -2x + 12$$

$$\frac{3y}{3} = \frac{-2x}{3} + \frac{12}{3} \qquad \text{Divide both sides by 3}$$

$$y = -\frac{2}{3}x + 4$$

$$m = -\frac{2}{3};\ b = 4$$

Writing equations in $y = mx + b$ form

Example 13:

Write the equation for the line with a slope of $\frac{3}{8}$ and a y-intercept of -11.

Solution:

$y = mx + b$ Substitute the slope of $\frac{3}{8}$ for m, and y-intercept of -11 for b

$y = \frac{3}{8}x - 11$

Writing an equation in $y = mx + b$ form from a graph

To write an equation for a linear function in slope-intercept form, all you need to know is the slope and the y-intercept.

Example 14:

Which function represents the graph below?

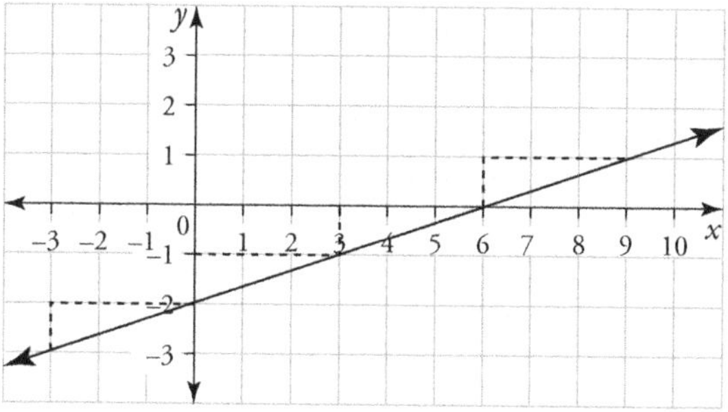

A. $y = 3x + 2$

B. $y = -3x - 2$

C. $y = \frac{1}{3}x - 2$

D. $y = 3x - 2$

Solution:

Choice C is the correct answer.

$$m = \frac{y_2 - y_1}{x_2 - x_1}$$

$$m = \frac{0-(-2)}{6-0}$$ Substitute $(0,-2)$ for (x_1, y_1) and $(6, 0)$ for (x_2, y_2)

$$m = \frac{2}{6}$$ Simplify

$$m = \frac{1}{3}$$

The graph has a slope of $\frac{1}{3}$ and a y-intercept of –2.

Graphing an equation in *y* = *mx* + *b* form

Example 15:

Graph an equation with a slope of $\frac{1}{5}$ and a y-intercept of 1.

Solution:

Step 1. Start by plotting the *y*-intercept, (0, 1).

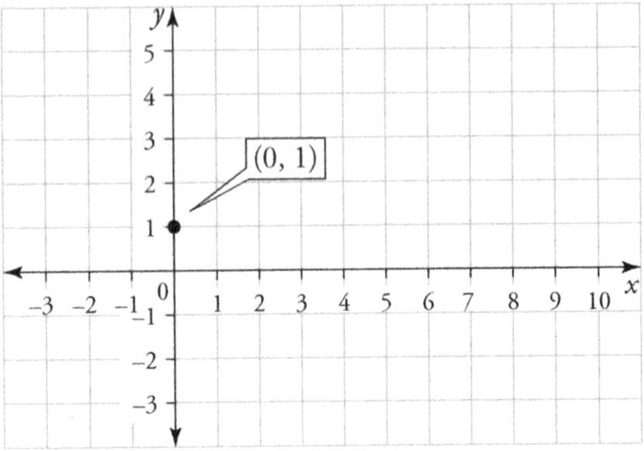

Step 2. Plot a second point using the slope $= \frac{\text{rise}}{\text{run}} = \frac{1}{5}$.

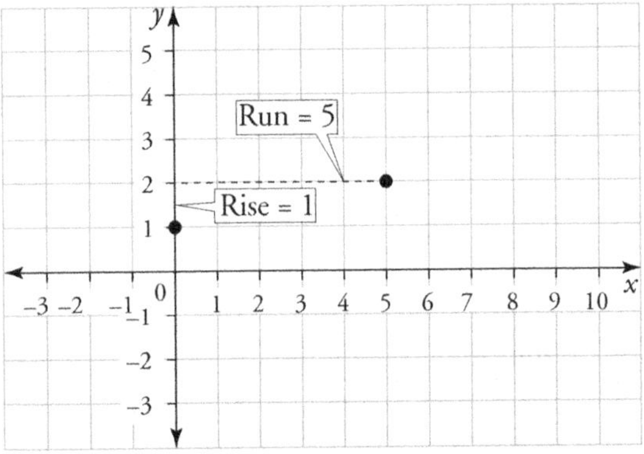

Step 3. Draw a line through the two points.

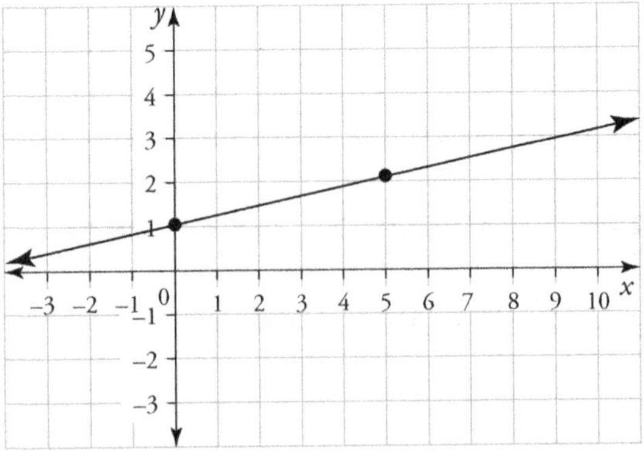

Example 16:

Graph the following equation using the slope and the y-intercept.

$$4x + 2y = 8$$

Solution:

Step 1. Isolate the variable y to rewrite the equation in $y = mx + b$ form.

$$\begin{aligned} 4x + 2y &= 8 && \\ -4x \quad\quad & -4x && \text{Subtract } 4x \text{ from both sides} \\ 2y &= -4x + 8 && \text{Simplify} \\ \frac{2y}{2} &= \frac{-4x}{2} + \frac{8}{2} && \text{Divide both sides by 2} \\ y &= -2x + 4 && \text{Simplify} \end{aligned}$$

Step 2. Identify the slope, m, and the y-intercept, b.

$$m = -2,\ b = 4$$

Step 3. Write the slope as a ratio of $\frac{\text{rise}}{\text{run}}$.

$$m = -2 = \frac{\text{rise}}{\text{run}} = \frac{-2}{1}$$

Step 4. Start by plotting the y-intercept, (0, 4).

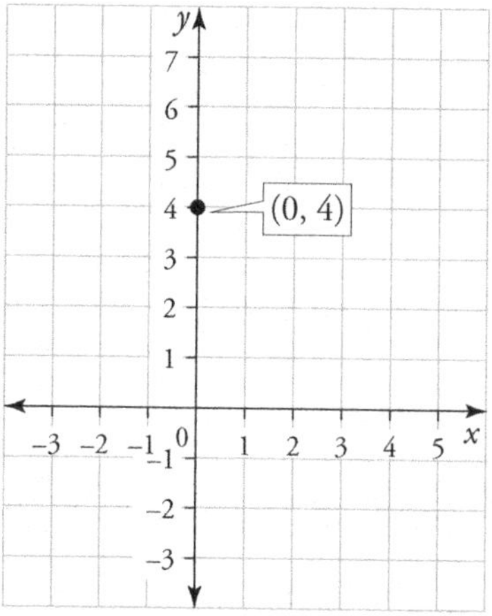

Step 5. Plot a second point using slope $= \frac{\text{rise}}{\text{run}} = \frac{-2}{1}$.

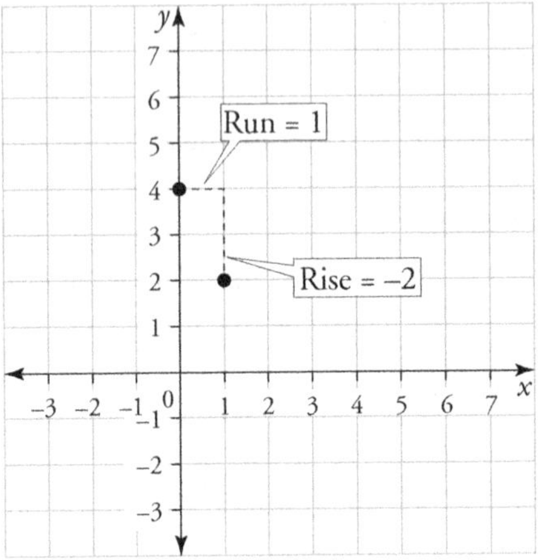

Note that the rise is negative, so to get from the *y*-intercept to the next point you move 1 unit to the right and 2 units down.

Step 6. Draw a line through the two points.

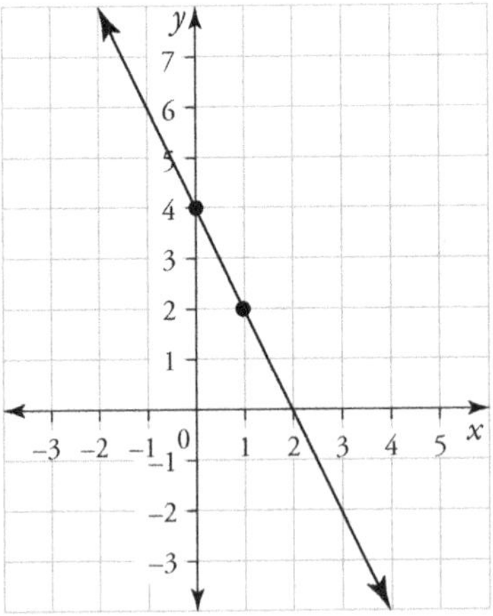

Point-Slope Form

You saw that you can graph a line by plotting a point on the *y*-intercept and using the slope to find other points on the graph. You can follow the same steps to graph a line as long as you know any point on the line and the slope of the equation.

Example 17:

Graph the line with a slope of $\frac{-1}{3}$ that passes through the point (2, 5).

Solution:

Step 1. Plot the point (2, 5).

Step 2. Plot a second point using the slope.

$$m = \frac{\text{rise}}{\text{run}} = \frac{-1}{3}$$

Move *down* 1 unit and 3 units to the *right.*

Step 3. Draw a line through the two points.

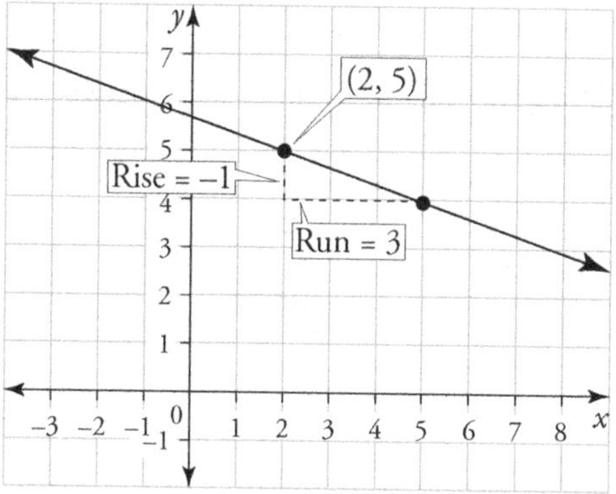

Writing equations in point-slope form

You can write the equation of any line if you know the slope and any point on that line using the **point-slope formula.**

Point-Slope Formula:

The equation of a line with slope m that passes through the point (x_1, y_1):

$$y - y_1 = m(x - x_1)$$

Example 18:

Write the point-slope equation of a line that has a slope of $\frac{1}{2}$ that passes through the point (5, –6).

Solution:

$$y - y_1 = m(x - x_1)$$

$$y - (-6) = \frac{1}{2}[x - (5)] \quad \text{Substitute: } x_1 = 5,\ y_1 = -6, \text{ and } m = \frac{1}{2}$$

$$y + 6 = \frac{1}{2}(x - 5) \quad \text{Simplify}$$

Writing equations from two points

Example 19:

Write an equation in slope-intercept form for the line that passes through (1, 3) and (7, –9).

Solution:

Step 1. Use the slope formula to find the slope between the two points.

$$\begin{aligned} m &= \frac{y_2 - y_1}{x_2 - x_1} \\ &= \frac{-9 - 3}{7 - 1} \\ &= \frac{-12}{6} \\ &= -2 \end{aligned}$$

Step 2. Substitute the slope and one of the points into point-slope form.

$$y - y_1 = m(x - x_1)$$

$$y - 3 = -2(x - 1) \quad \text{Substitute } (1, 3) \text{ for } (x_1, y_1)$$

Step 3. Isolate the y variable to rewrite as $y = mx + b$.

$$y - 3 = -2(x - 1) \quad \text{Distributive property}$$

$$y - 3 = -2x + 2 \quad \text{Simplify}$$

$$+3 \qquad +3 \quad \text{Add 3 to both sides.}$$

$$y = -2x + 5$$

Parallel Lines

Parallel lines are lines on a plane that never intersect.

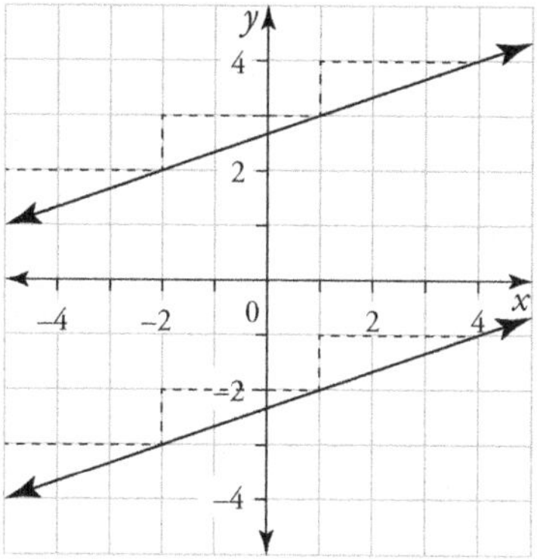

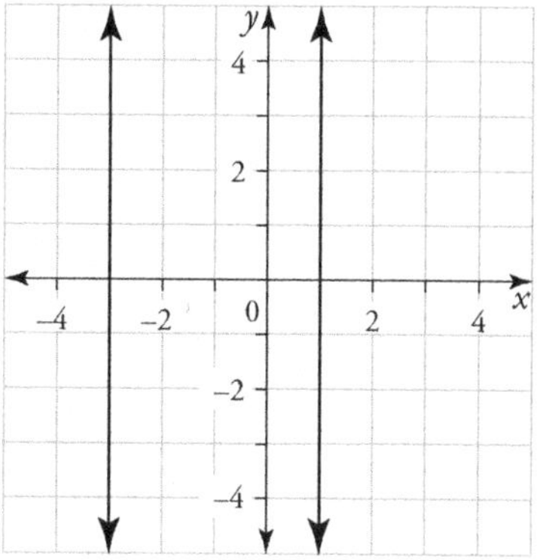

In the graph on the left, notice that both lines have a slope of $\frac{1}{3}$ but different y-intercepts.

> **Properties of Parallel Lines:**
>
> Parallel lines have the *same slope* but different y-intercepts. All vertical lines are parallel to the y-axis, and all horizontal lines are parallel to the x-axis.

Example 20:

Identify all of the parallel lines.

I. $y = -\frac{2}{3}x + 4$

II. $2x + 3y = 6$

III. $y - 4 = 3(x - 2)$

IV. $y = -\frac{2}{3}x - 7$

A. I and II

B. III and IV

C. I, II, and IV

D. All of the above

Solution:

Rewrite the equations in items II and III in slope-intercept form to compare slopes.

$$
\begin{aligned}
2x+3y&=6\\
-2x\qquad &-2x\\
3y&=-2x+6\\
\frac{3y}{3}&=\frac{-2x}{3}+\frac{6}{3}\\
y&=-\frac{2}{3}x+2\\
m&=-\frac{2}{3}
\end{aligned}
\qquad\qquad
\begin{aligned}
y-4&=3(x-2)\\
y-4&=3x-6\\
+4\qquad &+4\\
y\qquad &=3x-2\\
m&=3
\end{aligned}
$$

Now you can easily compare the slopes for each item:

I. $m=-\frac{2}{3}$ II. $m=-\frac{2}{3}$ III. $m=3$ IV. $m=-\frac{2}{3}$

The correct answer is C. Equations I, II, and IV all have a slope of $-\frac{2}{3}$.

Writing equations of parallel lines

Example 21:

Write the equation of a line that's parallel to $y = 7x + 1$ and passes through the point $(1, -6)$.

Solution:

Step 1. Parallel lines have the same slope. Identify the slope of $y = 7x + 1$.

$$m = 7$$

Step 2. Parallel lines have different y-intercepts. Rewrite the equation, leaving the y-intercept as the unknown variable, b.

$$y = 7x + b$$

Step 3. Substitute the coordinates of the point into x and y in the equation.

$$(-6) = 7(1) + b$$

Step 4. Solve for the y-intercept, b.

$$(-6) = 7(1) + b$$

$$(-6) = 7 + b \quad \text{Simplify}$$

$$-7 \quad -7 \quad \text{Subtract 7 from both sides}$$

$$-13 = b$$

Step 5. Substitute the new y-intercept for b and keep the same slope.

$$y = 7x - 13$$

Perpendicular Lines

Perpendicular lines are lines that intersect at right angles. Horizontal lines are perpendicular to vertical lines.

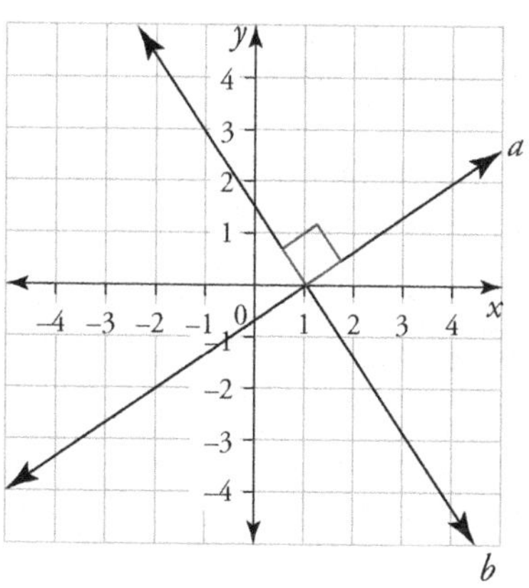

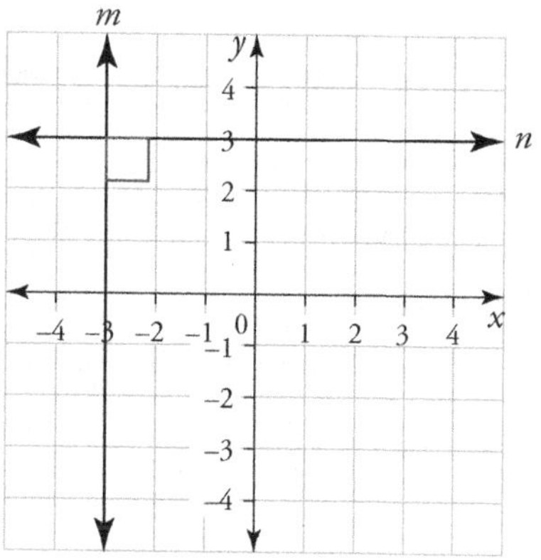

slope line $a = \frac{2}{3}$

slope line $b = -\frac{3}{2}$

slope line m = undefined

slope line $n = 0$

Properties of Perpendicular Lines:

Non-vertical lines are perpendicular if their slopes are negative reciprocals of each other. The product of their slopes will always equal –1.

Vertical and horizontal lines are always perpendicular to each other.

Example 22:

Given the slope of a line, find the slope of a line perpendicular to it.

A. $\frac{5}{7}$

B. –3

Solution:

	Slope	*Reciprocal*	*Negative Reciprocal*
A.	$\frac{5}{7}$	$\frac{7}{5}$	$-\frac{7}{5}$
B.	–3	$\frac{-1}{3}$	$\frac{1}{3}$

Writing equations of perpendicular lines

Example 23:

Write the equation of a line that's perpendicular to $4x + 2y = 18$ and passes through the point (–4, 9). Write in slope-intercept form.

Solution:

Step 1. Isolate the variable y to rewrite the equation as $y = mx + b$.

$$4x+2y=18$$

$$-4x \qquad -4x \qquad \text{Subtract } 4x \text{ from both sides}$$

$$2y=-4x+18 \qquad \text{Simplify}$$

$$\frac{2y}{2}=\frac{-4x}{2}+\frac{18}{2} \qquad \text{Divide both sides by 2}$$

$$y=-2x+9 \qquad \text{Simplify}$$

Step 2. Perpendicular lines have slopes that are negative reciprocals. Identify the slope of $y = -2x + 9$.

$$m = -2$$

Step 3. Rewrite the slope as its negative reciprocal.

$$-2 \Rightarrow \frac{1}{2}$$

Step 4. Substitute the coordinates of the point for x and y in the equation. Substitute the new slope.

$$(9) = \frac{1}{2}(-4) + b$$

Step 5. Solve for the y-intercept, b.

$$9 = \frac{1}{2}(-4) + b$$

$$9 = -2 + b \qquad \text{Multiply}$$

$$+2 = +2 \qquad \text{Add 2 to both sides}$$

$$11 = b$$

Step 6. Substitute the new y-intercept for b using the negative reciprocal as the slope.

$$y = \frac{1}{2}x + 11$$

Standard Form

There are certain situations in which both variables are equally dependent on each other. These relations are usually written in **standard form.**

Standard Form of a Linear Equation:

$Ax + By = C$, where A, B, and C are all integers, $A \neq 0$ and $B \neq 0$.

Finding *x*- and *y*-intercepts

To create the graph of a function in standard form, you will need to find the *x*- and *y*-intercepts.

***X*-intercepts,** also known as **zeros,** are points $(x, 0)$ at which a graph intersects the *x*-axis.

***Y*-intercepts** are points $(0, y)$ at which a graph intersects the *y*-axis.

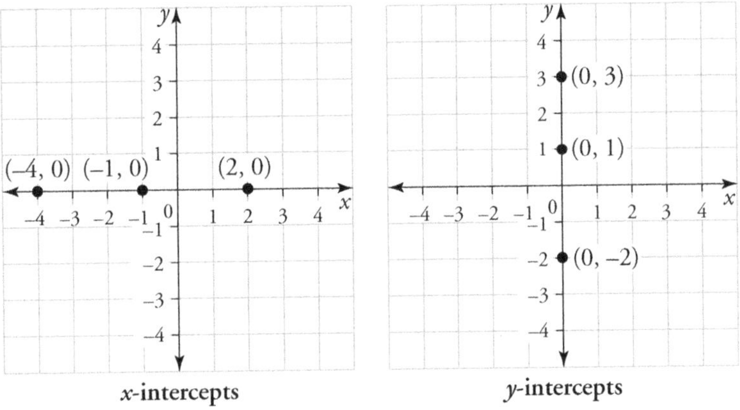

Example 24:

The snack counter at your school carnival is charging \$2 for hot dogs and \$3 for hamburgers. Between you and your friends, you have \$18 to spend. You want to find out all the different combinations of hot dogs, x, and hamburgers, y, you can purchase while spending exactly \$18. Write an equation to determine this.

Solution:

You can represent this relationship with the equation $2x + 3y = 18$.

Graphing equations in standard form

Example 25:

Graph the function $5x + 3y = 15$.

Solution:

Step 1. Find the x-intercept.

$$5x+3(0)=15 \quad \text{Substitute 0 for } y \text{ and solve for } x$$
$$5x=15 \quad \text{Simplify}$$
$$\frac{5x}{5}=\frac{15}{5} \quad \text{Divide both sides by 5}$$
$$x=3$$

Step 2. Find the y-intercept.

$$5(0)+3y=15 \quad \text{Substitute 0 for } x \text{ and solve for } y$$
$$3y=15 \quad \text{Simplify}$$
$$\frac{3y}{3}=\frac{15}{3} \quad \text{Divide both sides by 3}$$
$$y=5$$

Step 3. Plot the x-and y-intercepts, and draw a line through both points.

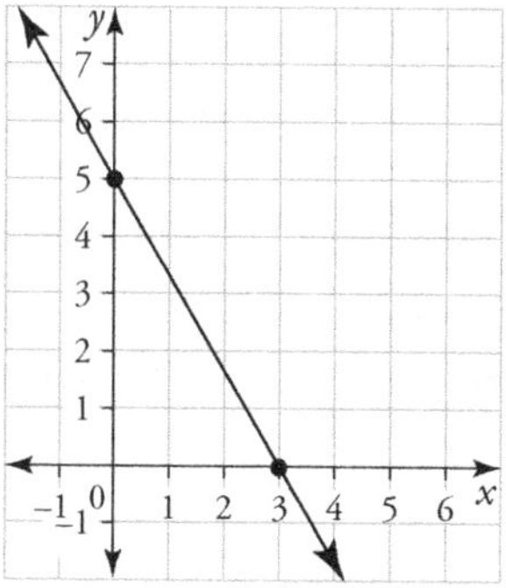

Equations of vertical and horizontal lines

Graphs of a vertical line and of a horizontal line are shown below.

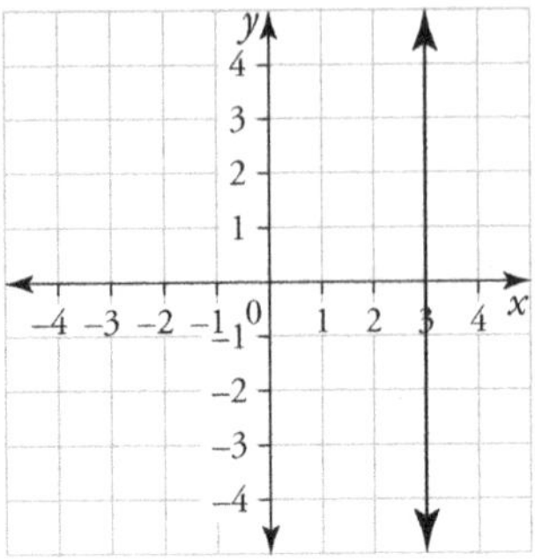

Vertical line

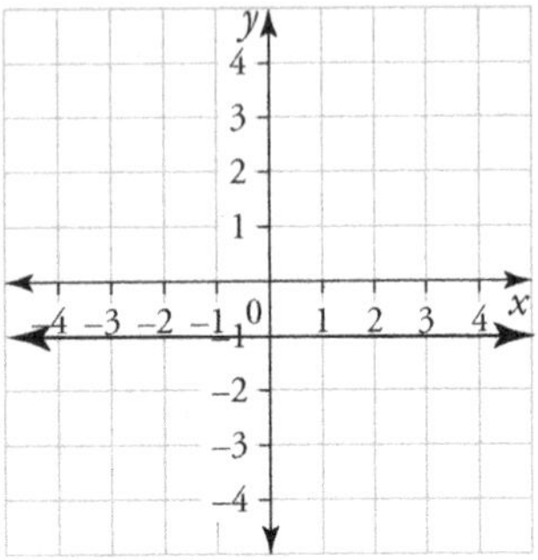

Horizontal line

Identify some points on each line.

x	***y***
3	–4
3	–2
3	0
3	1
3	3
For every value of *y*, *x* is always 3.	
Equation: $x = 3$	

x	***y***
–4	–1
–2	–1
0	–1
3	–1
5	–1
For every value of *x*, *y* is always –1.	
Equation: $y = -1$	

Equations of Horizontal Lines:

$y = b$, where b represents the y-intercept of the horizontal line.

Equations of Vertical Lines:

$x = b$, where b represents the x-intercept of the vertical line.

Example 26:

Graph the following equations.

$x = -4$ $\qquad$ $y = 2$ $\qquad$ $x = 0$

Solution:

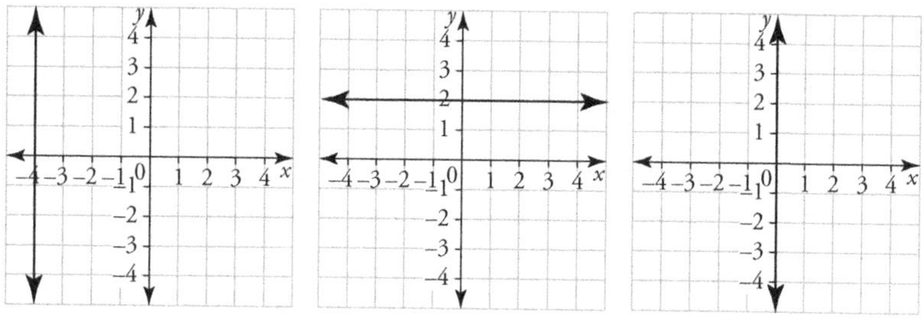

Transforming linear equations into standard form

Example 27:

Write $y = -\frac{1}{3}x + 4$ in standard form using integers.

Solution:

Step 1. Multiply each side by the denominator of the fractional slope to turn fractions into integers.

$$3(y) = 3\left(-\frac{1}{3}x + 4\right) \quad \text{Multiply both sides by 3}$$

Step 2. Isolate the constant to rewrite as $Ax + By = C$.

$3y = -x + 12$ Simplify using distributive property

$+x \quad +x$ Add $1x$ to both sides

$x + 3y = 12$ Simplify

Example 28:

Write $y = \frac{7}{4}x - 7$ in standard form.

Solution:

Step 1. Multiply each side by the denominator of the fractional slope to turn fractions into integers.

$$4(y) = 4\left(\frac{7}{4}x - 7\right) \quad \text{Multiply both sides by 4}$$

Step 2. Isolate the constant to rewrite as $Ax + By = C$.

$$\begin{aligned} 4y &= 7x - 28 && \text{Simplify using distributive property} \\ -7x &= -7x && \text{Subtract } 7x \text{ from both sides} \\ -7x + 4y &= -28 && \text{Simplify} \end{aligned}$$

Step 3. Make the constant positive.

$$\begin{aligned} -1(-7x + 4y) &= -1(-28) && \text{Multiply both sides by } -1 \\ 7x - 4y &= 28 && \text{Simplify} \end{aligned}$$

Linear Parent Function

Recall that a **parent function** is the simplest form of a function.

> **Linear Parent Function:**
>
> $y = x$ or $f(x) = x$

Here is a graph of the linear parent function:

x	f(x)
–2	–2
–1	–1
0	0
1	1
2	2

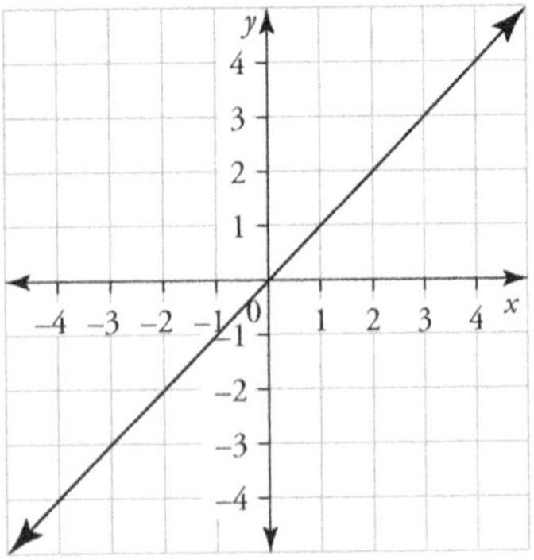

Effects on graph of linear parent function

Families of functions have the same defining characteristics as their parent function.

Let's explore the effects on the graph of the parent function when the independent and dependent variables are changed.

Effects of $f(x) + d$

First, compare the effects on the graph of the parent function when you add a value, d, to the dependent variable, $f(x)$.

	Function *f*	**Function *g***	**Function *h***
verbal description	Parent function	add 4 to $f(x)$	subtract 3 from $f(x)$
equation	$y = x$	$y = x + 4$	$y = x - 3$
equation in function notation	$f(x) = x$	$g(x) = f(x) + 4$	$h(x) = f(x) - 3$

x	***f(x)***	**(*x*, *f(x)*)**
–1	**–1**	(–1, –1)
0	**0**	(0, 0)
1	**1**	(1, 1)
2	**2**	(2, 2)

x	***f(x)* + 4**	**(*x*, *g(x)*)**
–1	**–1** + 4 = 3	(–1, 3)
0	**0** + 4 = 4	(0, 4)
1	**1** + 4 = 5	(1, 5)
2	**2** + 4 = 6	(2, 6)

x	***f(x)* – 3**	**(*x*, *h(x)*)**
–1	**–1** – 3 = –4	(–1, –4)
0	**0** – 3 = –3	(0, –3)
1	**1** – 3 = –2	(1, –2)
2	**2** – 3 = –1	(2, –1)

Here are graphs of $f(x) + d$:

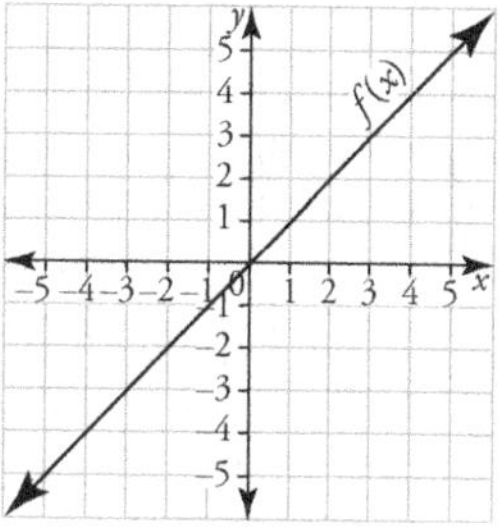

$f(x)$
$y = x$
Parent function

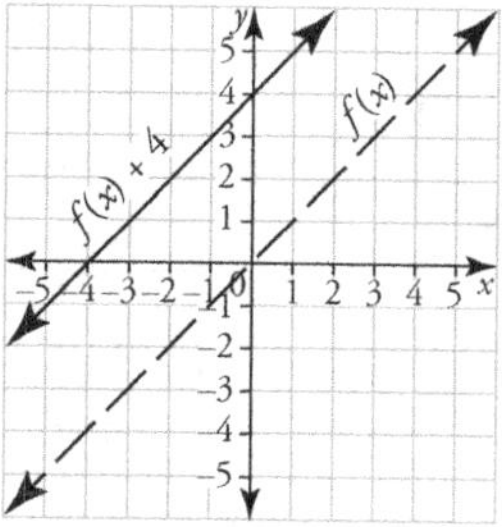

$f(x) + 4$
$y = x + 4$
Translated up 4 units

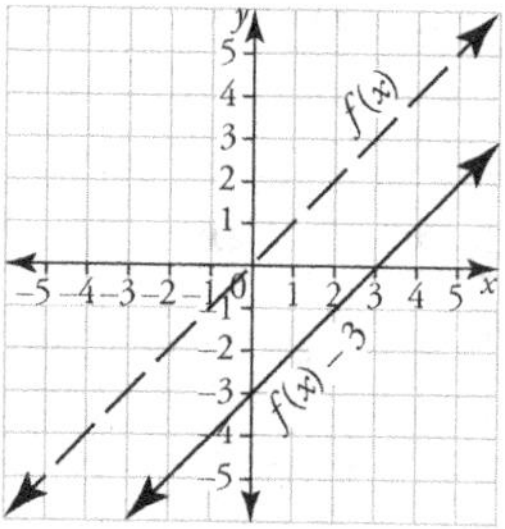

$f(x) - 3$
$y = x - 3$
Translated down 3 units

Effects of $f(x) + d$:

If $y = f(x)$, then $y = f(x) + d$ results in a vertical translation of "d units up" when d is positive or "d units down" when d is negative.

Example 29:

If the graph of $y = -3x + 2$ is translated as 6 units down, which equation describes the new graph?

A. $y = 3x + 2$
B. $y = -3x - 4$
C. $y = -3x + 8$
D. $y = -9x + 2$

Solution:

The correct answer is B. Subtracting 6 from x results in a vertical translation of down 6 units.

Effects of *f (bx)*

Now let's compare the effects on the graph of the parent function when you multiply the independent variable, x, by a value, b.

multiply x by 2

$f(2x)$ or $y = 2x$

x	**2x**	**(x, f(2x))**
−1	$2(-1) = -2$	$(-1, -2)$
0	$2(0) = 0$	$(0, 0)$
1	$2(1) = 2$	$(1, 2)$
2	$2(2) = 4$	$(2, 4)$

multiply x by $\frac{1}{4}$

$f\left(\frac{1}{4}x\right)$ or $y = \frac{1}{4}x$

x	$\frac{1}{4}x$	$\left(x, f\left(\frac{1}{4}x\right)\right)$
−1	$\frac{1}{4}(-1) = -\frac{1}{4}$	$\left(-1, -\frac{1}{4}\right)$
0	$\frac{1}{4}(0) = 0$	$(0, 0)$
1	$\frac{1}{4}(1) = \frac{1}{4}$	$\left(1, \frac{1}{4}\right)$
2	$\frac{1}{4}(2) = \frac{1}{2}$	$\left(2, \frac{1}{2}\right)$

Here are graphs of $f(bx)$:

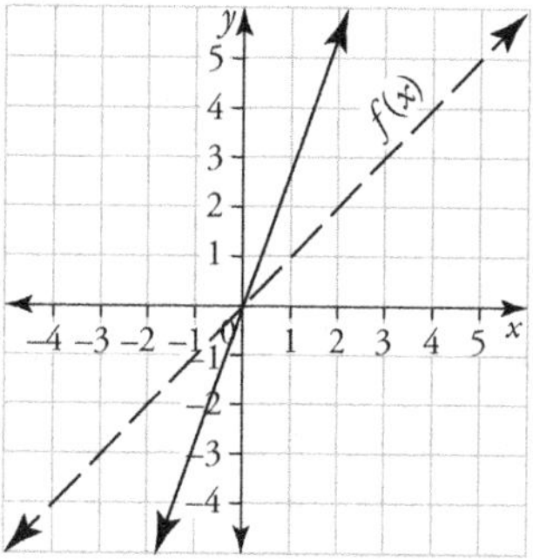

$f(2x)$

$y = 2x$

Slope is doubled;
line is steeper

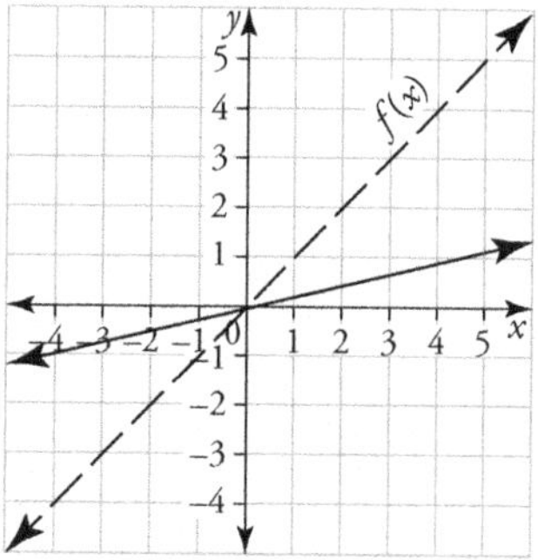

$f\left(\frac{1}{4}x\right)$

$y = \frac{1}{4}x$

Slope is $\frac{1}{4}$ of original slope;
line is less steep

Observe the effects on the graph of the parent function when b is negative.

Here is a graph of $y = -x$:

Multiply x by -1

$f(-x)$ or $y = -x$

x	**–x**	**(x, f(–x))**
–2	–(–2) = 2	(–2, 2)
–1	–(–1) = 1	(–1, 1)
0	–(0) = 0	(0, 0)
1	–(1) = –1	(1, –1)
2	–(2) = –2	(2, –2)

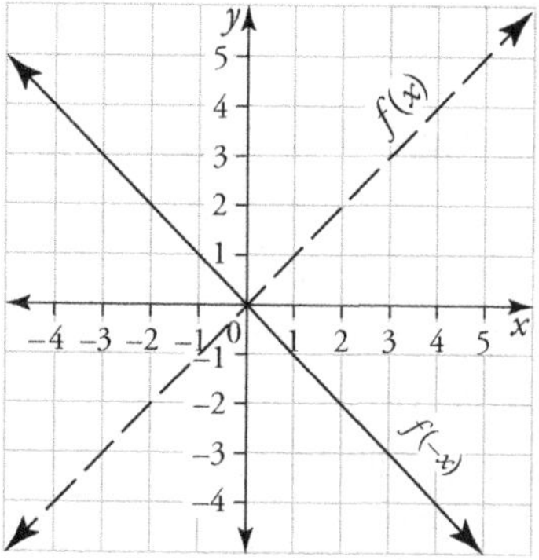

Slope becomes negative;
line falls from left to right

Example 30:

Two functions are given below.

$$g(x) = 4x + 9$$
$$h(x) = -4x + 9$$

How does the graph of function *h* compare to the graph of function *g*?

A. The graph of *h* has the same *y*-intercept as the graph of *g*.
B. The graph of *h* is translated 8 units down from *g*.
C. The graph of *h* is steeper than the graph of *g*.
D. The graph of *h* is perpendicular to the graph of *g*.

Solution:

The correct answer is A. The graphs of $g(x)$ and $h(x)$ have different slopes but the same *y*-intercept.

Effects of $f(bx)$ in a Linear Function:

If $y = f(x)$ is a linear function, then $f(bx)$ results in a change in the slope of the line.

Scatter Plots

A **scatter plot** is a graph that shows the relationship between two variables. Data is collected and plotted on a coordinate plane in order to establish if a relationship exists.

Example 31:

A teacher polls the class on how many hours they spend studying for their math midterm test. He then compares their responses with their actual scores in the table below. Make a scatter plot showing the relationship between the number of hours a student spends studying for the math test and his or her actual score on the math test.

Solution:

Let *x* represent the number of hours spent studying, and let *y* represent the student's test score.

x	y
3	100
1	71
1	78
$\frac{1}{2}$	58
0	60
4	94
3	90

x	y
2	84
$1\frac{1}{2}$	80
2	80
$2\frac{1}{2}$	90
3	96
1	64
$\frac{1}{2}$	62

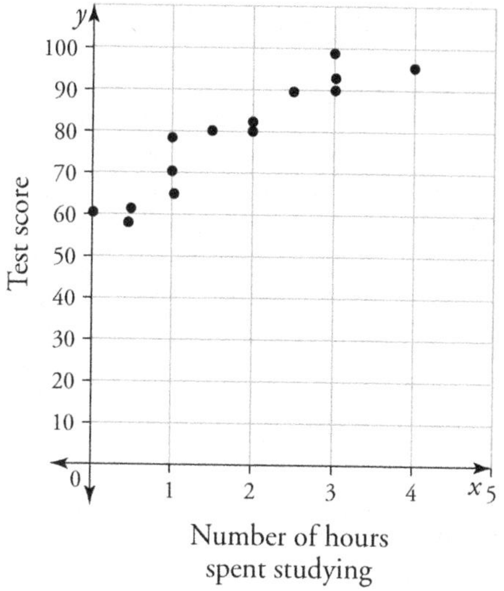

Notice that, in general, the more hours a student spent studying, the higher he or she scored on the test. We can say there is a **positive correlation** between the number of hours a student spends studying and his or her test score.

Types of correlations

Positive correlation In general, as one data value increases, the other value also increases.	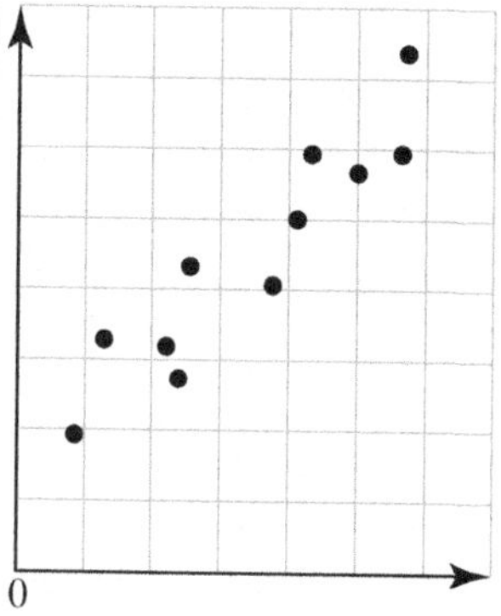

Negative correlation In general, as one set of data increases, the other decreases.	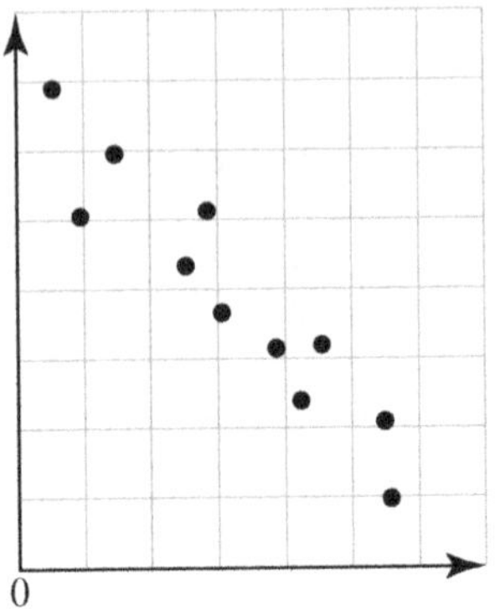
No correlation No apparent relationship exists between the two sets of data.	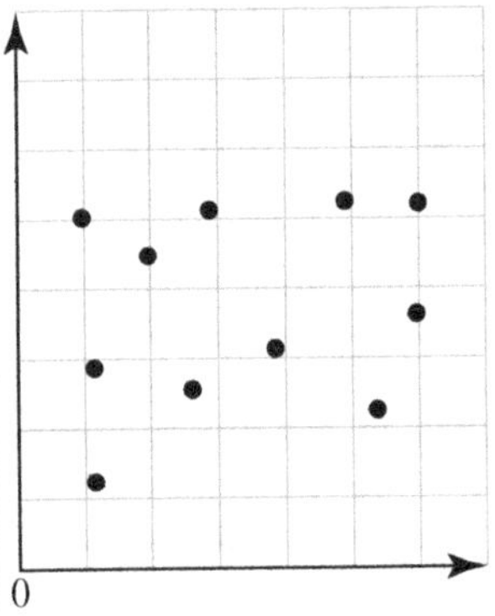

Example 32:

In the situations below, indicate what type of correlation you would expect between the two sets of data. Explain.

A. The number of calories burned and the amount of time you spend sitting down

B. The ages of students at elementary, middle, and high schools and their height

C. The outdoor temperature and the amount of electricity used by the heater

Solution:

A. Negative: The more time you spend sitting down, the fewer calories you burn; as one value increases, the other decreases.

B. Positive: In general, as students from grades K–12 get older, the taller they get; as one value increases, so does the other.

C. Negative: The higher the outdoor temperature becomes, the less time a heater will be used.

TECH TIPS: Creating Scatter Plots on a Graphing Calculator

Creating a scatter plot from data

Example:

The weighing of wild boars for scientific purposes can be a difficult process. In their natural habitat, they are difficult to capture and must be sedated, which can become costly. A team of scientists is contracted to create a system for estimating their weight based on their body length from an aerial picture. Make a scatter plot of the following data collected in order to establish the relationship between a wild boar's body length and its weight.

Length (in)	40	60	39	77	35	41	58	79	58	50	71	65	48	44
Weight (lb)	112	184	121	189	110	142	166	200	185	173	194	177	158	135

Solution:

1. Press STAT.

   Select option **1: Edit.**

EDIT	CALC	TESTS
1: Edit...		
2: SortA(		
3: SortD(		
4: ClrList		
5: SetUpEditor		

2. Enter input values under **L1.**

   Press ENTER to enter next value.

L1	L2	L3
40	112	
60	184	
39	121	
77	189	
35	110	
41	142	
58	166	
L1(1)=40		

3. Enter output values under **L 2** next to corresponding input values.

4. Press 2nd Y= to access **STAT PLOTS** menu.

   Select option **1: Plot 1.**

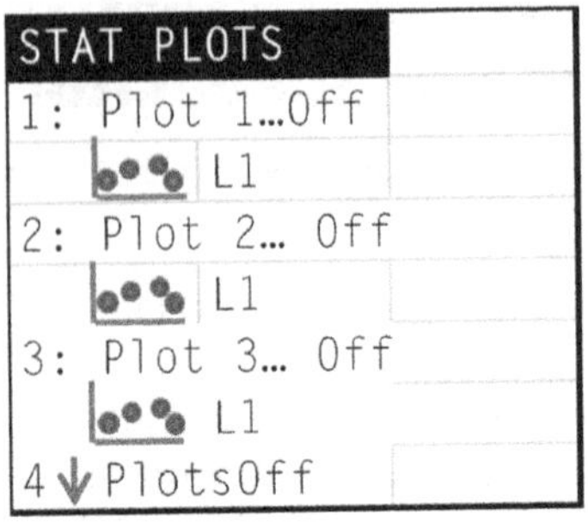

5. Turn Plot 1 on by highlighting **On** and press ENTER.

   Leave **Type:** on scatter plot and settings as shown.

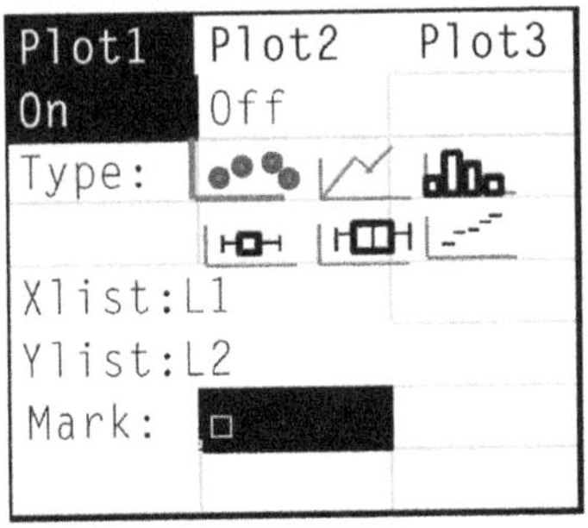

6. Press WINDOW to change window settings to fit the data as shown.

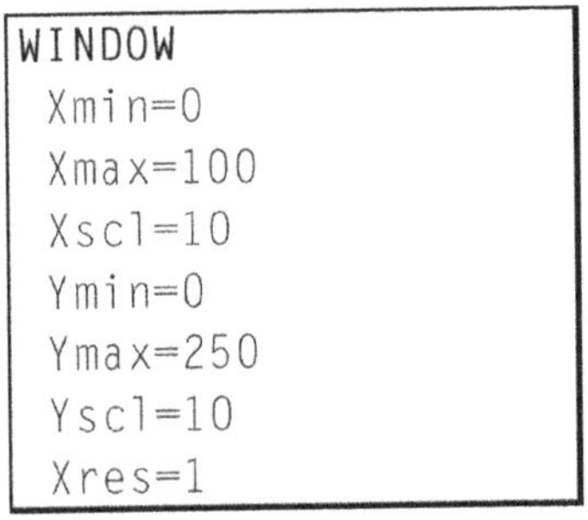

7. Press GRAPH to view scatter plot.

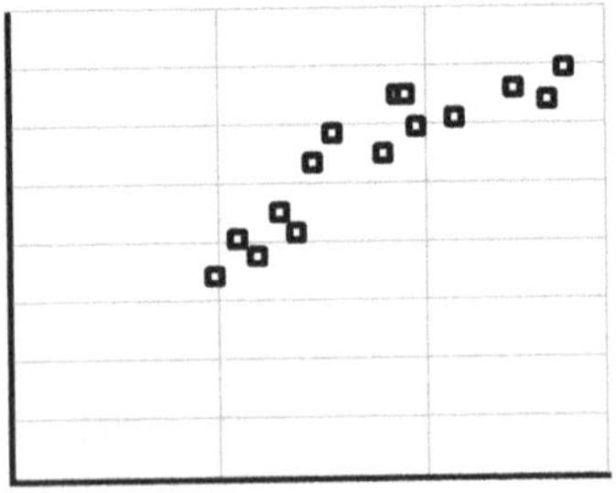

Finding the line of best fit

Example:

The **line of best fit** is a line that shows the general trend for a specific group of points on a scatter plot. Your graphing calculator can use linear regression to find the line of best fit.

Solution:

1. After entering all your data points using **STAT PLOT**, press **STAT**.

   Press the right arrow key to highlight the **CALC** menu.

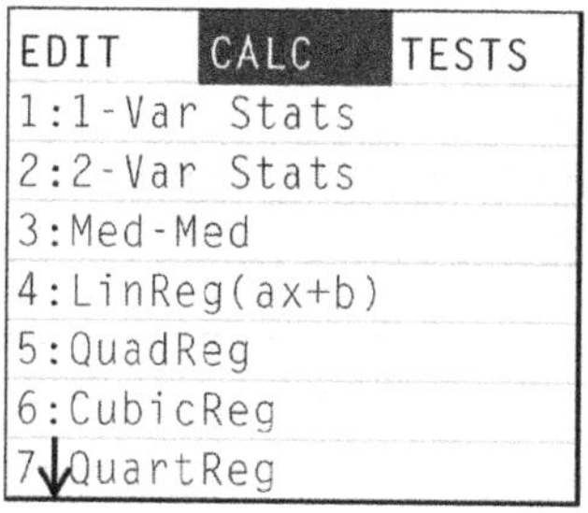

2. Select option **4: LinReg (ax+b).**

3. Once in the home screen, press **ENTER**.

   The equation for the line of best fit will be displayed.

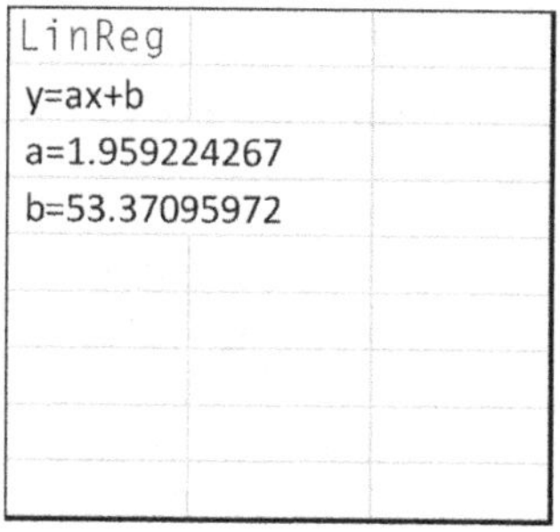

4. Press **Y =** .

   Enter the equation for the line of best fit in Y_1.

   $y = 1.96x + 53.37$

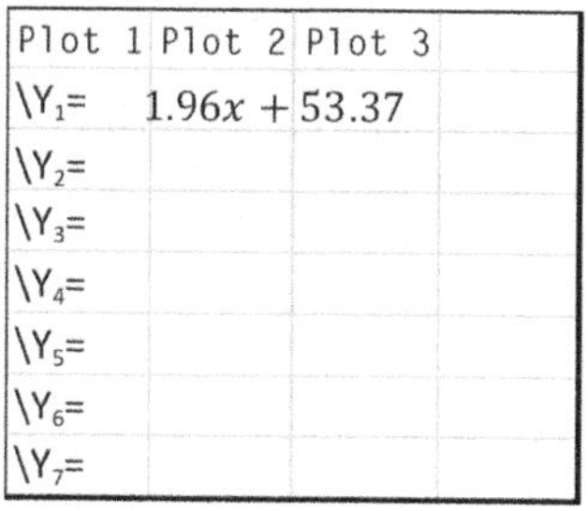

5. Press **GRAPH**.

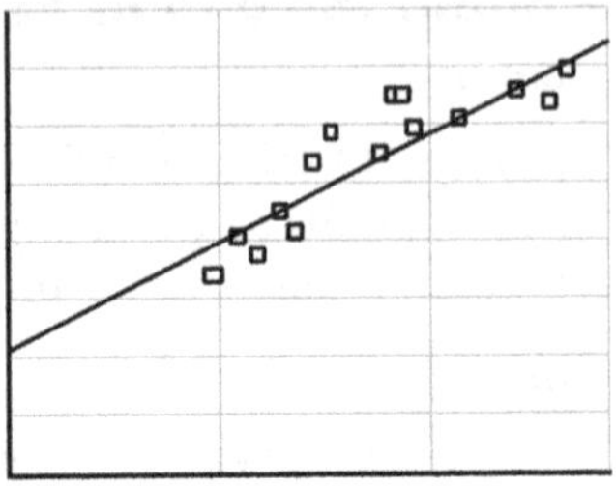

Finding the correlation coefficient

Example:

The **correlation coefficient,** r, is a numerical value between –1 and 1 that describes how well your line of best fit actually represents your data.

If your data is perfectly linear, the correlation coefficient $r = \pm 1$.

If r is close to +1, this indicates a strong positive correlation.

If r is close to –1, this indicates a strong negative correlation.

If $r = 0$, this indicates there is no correlation between the variables.

Solution:

1. Turn on Diagnostics On the Catalog menu by pressing **2nd** **0**.

   Arrow down to Diagnostic On and press **ENTER** twice.

2. Follow steps 1–3 above for finding the line of best fit. The correlation coefficient, r, will appear on the screen.

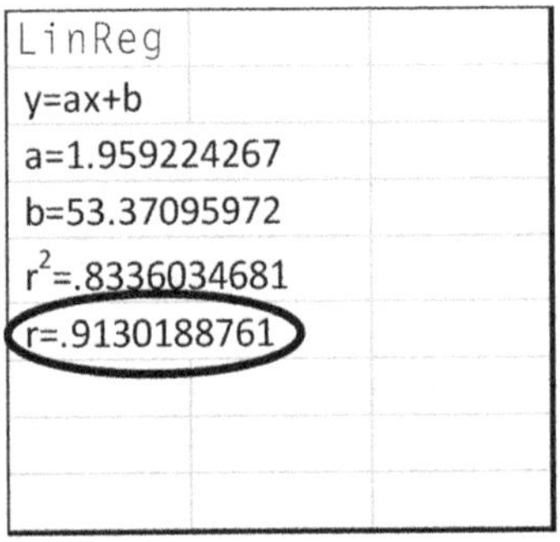

Chapter Check-Out

Questions

For questions 1–2, calculate the rate of change and explain what it means in each situation.

1.

Number of Miles	*Calories Burned while Running*
2	248
3	372
5	620
6	744

2.

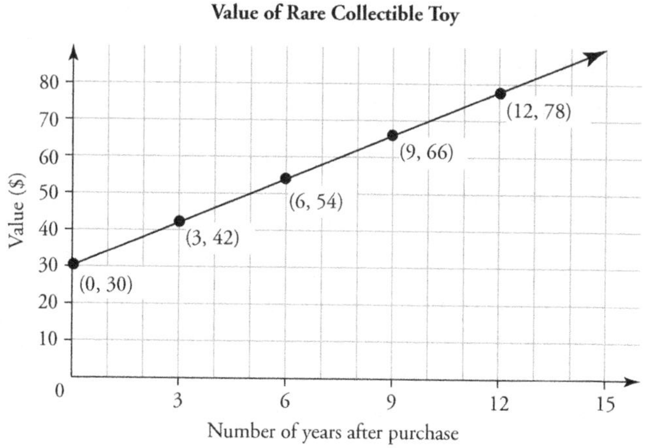

For questions 3–4, find the slope of each line.

3.

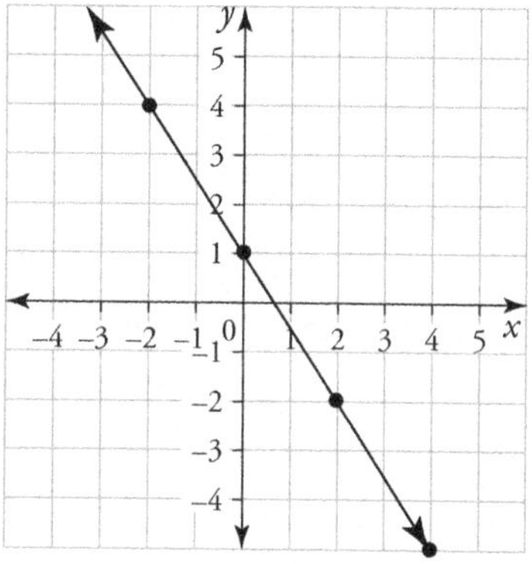

4.

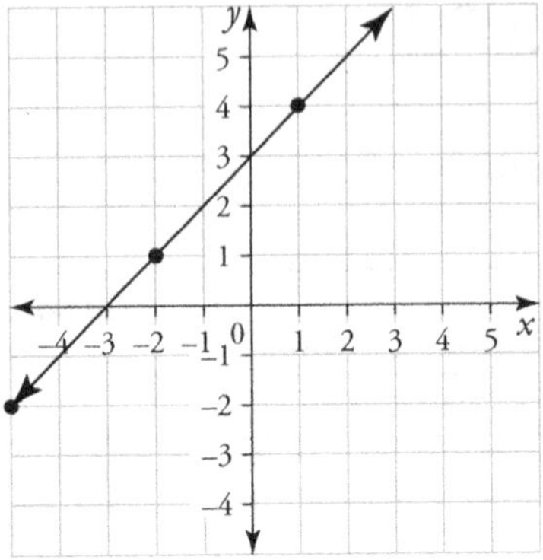

For questions 5–6, find the slope of the line that contains the following points.

5. (0, 13) and (9, 20)

6. (–5, 15) and (6, –7)

For questions 7–8, determine if each equation is a direct variation. If so, find the constant of variation, *k*.

7. $3x + 4y = 0$

8. $2x - 5y = 10$

For questions 9–11, write the equation of the line in slope-intercept form for:

9. A line with a slope of 3 and a *y*-intercept of –5.

10. A line that passes through the points (–5, 0) and (10, 3).

11. The line shown below.

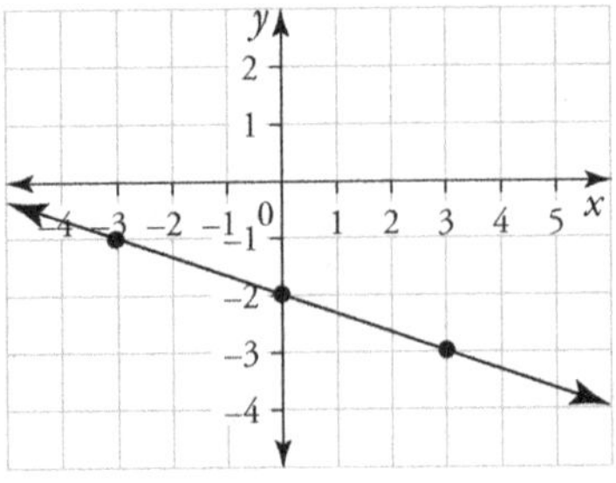

For questions 12–13, write an equation in point-slope form for:

12. A line with a slope of 8 that passes through the point (–1, 3).

13. A line that passes through the points (4, –22) and (–1, 8).

For questions 14–15, write an equation of the line with the following description. Write in slope-intercept form.

14. A line parallel to $y = 2x + 6$ and passes through the point (4, –6).

15. A line perpendicular to $y = \frac{5}{3}x - 1$ and passes through the point (–5, 7).

For questions 16–17, find the x- and y-intercepts of the following functions.

16. $4x - 6y = 8$

17. $-3x - 7y = 21$

For questions 18–19, determine the effects on the graph of the linear parent function, $y = x$.

18. $y = x - 11$

19. $y = -4x$

For questions 20–21, determine the effects on the graph of the function $f(x) = 2x + 3$.

20. $f\left(\frac{1}{7}x\right)$

21. $f(x) + 6$

Answers

1. 124 calories burned per mile
2. The value of this rare collectible toy increases \$4 per year.
3. $-\frac{3}{2}$
4. 1
5. $\frac{7}{9}$
6. –2
7. yes; $k = -\frac{3}{4}$
8. not a direct variation
9. $y = 3x - 5$
10. $y = \frac{1}{5}x + 1$
11. $y = -\frac{1}{3}x - 2$
12. $y - 3 = 8(x + 1)$
13. $y + 22 = -6(x - 4)$ or $y - 8 = -6(x + 1)$
14. $y = 2x - 14$
15. $y = -\frac{3}{5}x + 4$
16. $(2,0)$ and $\left(0, -\frac{4}{3}\right)$
17. $(-7, 0)$ and $(0, -3)$
18. vertical shift 11 units down
19. slope becomes steeper and falls from left to right
20. slope becomes less steep
21. vertical shift 6 units up

Chapter 6

SYSTEMS OF LINEAR EQUATIONS

Chapter Check-In

- ❑ Solving systems of linear equations by graphing
- ❑ Solving systems of linear equations by substitution
- ❑ Solving systems of linear equations by elimination
- ❑ Types of solutions to systems of linear equations
- ❑ Linear inequalities
- ❑ Systems of linear inequalities
- ❑ TECH TIPS: Solving systems of linear equations and graphing linear inequalities on a graphing calculator

Two or more linear equations representing the same two variables are known as a **system of linear equations.** The **solution to a system of linear equations** is any point that makes all equations in the system true.

If a certain ordered pair makes more than one equation true, then the graphs of those equations will share that point in common, and the graphs of those equations will intersect at that point.

Solving Systems of Linear Equations by Graphing

Example 1:

Your school's student council is selling t-shirts with the school logo to raise money for prom. *Carol's Custom T-shirts* charges \$130 to set up the template for your school logo and an additional \$5 for every t-shirt they print. You plan on selling the t-shirts for \$10 each.

Write and solve a system of equations that will allow you to find the number of t-shirts, x, the student council needs to order and sell in order to break even.

Solution:

Verbal	Equation
Money spent = \$130 plus \$5 per t-shirt printed	$y = 130 + 5x$
Money earned = \$10 per t-shirt sold	$y = 10x$

To find the number of t-shirts you need to print and sell in order to break even, the amount of money spent must equal the amount of money earned.

Graph both equations on the same coordinate plane and find the **point of intersection.**

$y = 130 + 5x$
y-intercept of 130
and slope of 5

$y = 10x$
y-intercept of 0
and slope of 10

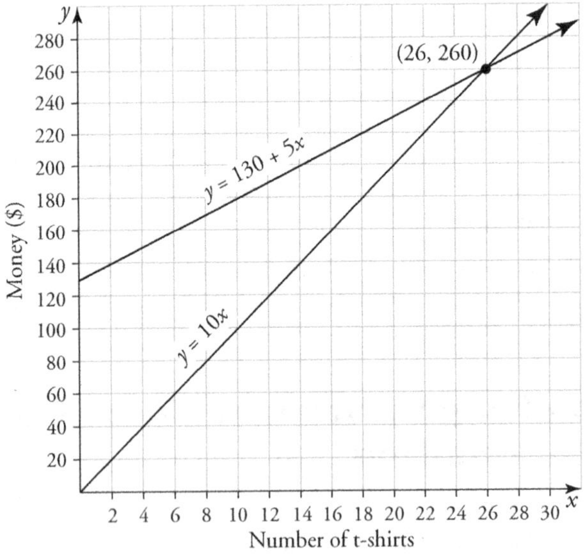

The lines intersect at (26, 260).

Check:

To check the solution, substitute 26 in place of x for both equations.

$y = 130 + 5x$ | $y = 10x$
$y = 130 + 5(26)$ | $y = 10(26)$
$y = 130 + 130$ | $y = 260$
$y = 260$

Check

If the student council purchases 26 t-shirts, they will spend $260.

If the student council sells those 26 t-shirts, they will earn $260 and break even.

Example 2:

Solve by graphing.

$$\begin{cases} y = 2x - 4 & y\text{-intercept of } -4\text{, slope of } 2 \\ y = -x + 2 & y\text{-intercept of } 2\text{, slope of } -1 \end{cases}$$

Solution:

Point of intersection (2, 0):

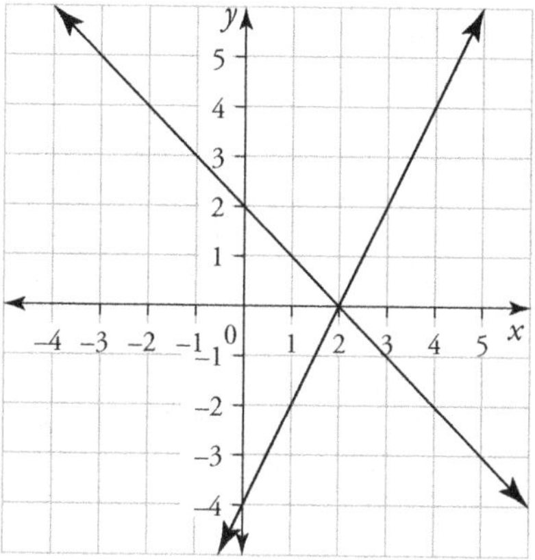

Check:

Substitute 2 in place of x in both equations.

$$\begin{array}{lcl} y = 2x - 4 & & y = -x + 2 \\ y = 2(2) - 4 & & y = -(2) + 2 \\ y = 0 & \checkmark & y = 0 \end{array}$$

Rewriting in slope-intercept form

You can rewrite equations in slope-intercept form to solve by graphing.

Example 3:

Solve by graphing.

$$\begin{cases} -x + 3y = 12 \\ y = -2x - 3 \end{cases}$$

Solution:

Rewrite the first equation in slope-intercept form, $y = mx + b$, to find the y-intercept and the slope.

$$\begin{array}{rl} -x + 3y = 12 & \\ +x \qquad\quad +x & \text{Add } x \text{ to both sides} \\ 3y = x + 12 & \text{Simplify} \\ \dfrac{3y}{3} = \dfrac{x}{3} + \dfrac{12}{3} & \text{Divide both sides by 3} \\ y = \dfrac{1}{3}x + 4 & \end{array}$$

$$\begin{cases} y = \dfrac{1}{3}x + 4 & y\text{-intercept of 4, slope of } \dfrac{1}{3} \\ y = -2x - 3 & y\text{-intercept of } -3\text{, slope of } -2 \end{cases}$$

Point of intersection (–3, 3):

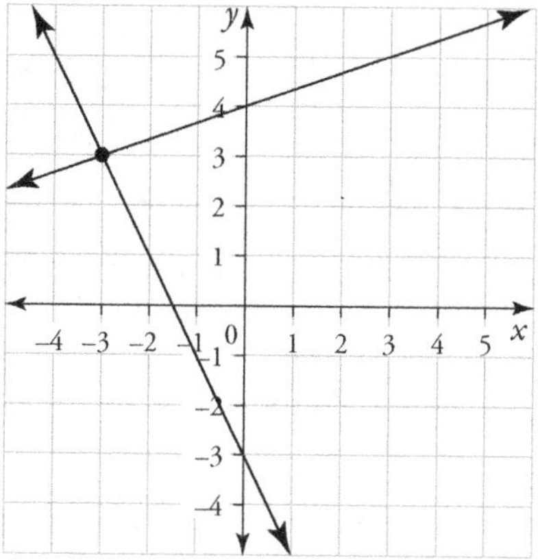

Check:

Substitute –3 in place of *x* for both equations.

$$y = \frac{1}{3}(-3) + 4 \qquad y = -2(-3) - 3$$
$$y = -1 + 4 \qquad y = 6 - 3$$
$$y = 3 \quad \checkmark \quad y = 3$$

Solving Systems of Linear Equations by Substitution

Not all systems of linear equations will have whole-number solutions. Sometimes graphs will intersect at points that are not easy to view on a graph. Another method for solving a system of linear equations is through **substitution.**

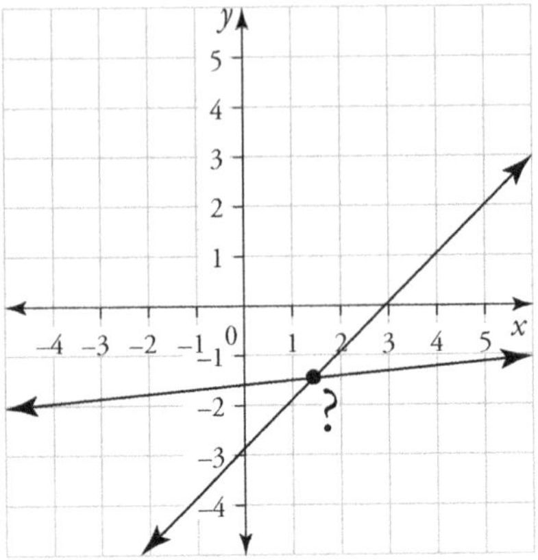

By substituting an expression in place of a variable, you can rewrite an equation using only one variable, allowing you to solve for that variable.

Example 4:

Solve by substitution.

$$\begin{cases} y = x + 1 \\ 3x + y = -4 \end{cases}$$

Solution:

Step 1. Substitute one expression into the other equation.

$y = x + 1$ → $3x + y = -4$ — Substitute $x + 1$ for y in the second equation

Step 2. Solve for the variable.

$$3x + x + 1 = -4$$

$4x + 1 = -4$ — Combine like terms

$-1 \quad -1$ — Subtract 1 from both sides

$4x = -5$ — Simplify

$\frac{4x}{4} = \frac{-5}{4}$ — Divide both sides by 4

$x = -1.25$ — Simplify

Step 3. Substitute the value of x into either equation and solve for y.

$$y=(-1.25)+1 \quad \text{Substitute} -1.25 \text{ for } x \text{ and solve for } y$$
$$y=-0.25 \quad \text{Simplify}$$

Step 4. Check by substituting both values into the other equation.

$$3(-1.25)+-0.25=-4 \quad \text{Substitute} -1.25 \text{ for } x \text{ and} -0.25 \text{ for } y$$
$$-3.75-0.25=-4 \quad \text{Simplify}$$
$$-4=-4 \quad \checkmark$$

Step 5. Write the solution as an ordered pair (x, y).

$$(-1.25, -0.25)$$

Example 5:

Solve by substitution.

$$\begin{cases} y=3x-8 \\ y=-5x+7 \end{cases}$$

Solution:

Step 1. Substitute one expression into the other equation.

$$y=3x-8$$
$$y=-5x+7$$

Substitute $3x - 8$ for y in the second equation

Step 2. Solve for the variable.

$3x-8=-5x+7$	
$+5x \qquad +5x$	Group variables on one side; add $5x$ to both sides
$8x-8=7$	Simplify
$+8 \quad +8$	Add 8 to both sides
$8x=15$	Simplify
$\frac{8x}{8}=\frac{15}{8}$	Divide both sides by 8
$x=1.875$	Simplify

Step 3. Substitute the value of x into either equation and solve for y.

$y=3(1.875)-8$	Substitute 1.875 for x and solve for y
$y=5.625-8$	Simplify (multiplication)
$y=-2.375$	Simplify (subtraction)

Step 4. Check by substituting both values into the other equation.

$-2.375=-5(1.875)+7$	Substitute 1.875 for x and -2.375 for y
$-2.375=-9.375+7$	Simplify (multiplication)
$-2.375=-2.375$	✓

Step 5. Write the solution as an ordered pair (x, y).

$$(1.875, -2.375)$$

Isolating a variable

Using substitution to solve a system can be convenient if one or both equations are already solved for a variable, as in Examples 2 and 3. If not, you can still use substitution by first isolating a variable in one of your equations.

Example 6:

Solve by substitution.

$$\begin{cases} 5x+2y=8 \\ x-4y=6 \end{cases}$$

Solution:

Step 1. Solve for one variable in either equation.

$x-4y=6$	
$+4y \quad\quad +4y$	Add $4y$ to both sides
$x=4y+6$	Simplify

Step 2. Substitute one expression into the other equation.

$x=(4y+6)$ $5(x)+2y=8$	Substitute $4y+6$ for x in the other equation

Step 3. Solve for the variable.

$5(4y+6)+2y=8$	
$20y+30+2y=8$	Distributive property
$22y+30=8$	Combine like terms
$-30 \quad -30$	Subtract 30 from both sides
$22y=-22$	Simplify
$\frac{22y}{22}=\frac{-22}{22}$	Divide both sides by 22
$y=-1$	Simplify

Step 4. Substitute the value of y into either equation and solve for x.

$x-4(-1)=6$	Substitute -1 for y and solve for x
$x+4=6$	Simplify (multiplication)
$-4 \quad -4$	Subtract 4 from both sides
$x=2$	Simplify

Step 5. Check by substituting both values into the other equation.

$$\begin{aligned} 5(2)+2(-1)&=8 && \text{Substitute 2 for } x \text{ and } -1 \text{ for } y \\ 10+(-2)&=8 && \text{Simplify (multiplication)} \\ 8&=8 && \checkmark \end{aligned}$$

Step 6. Write the solution as an ordered pair (x, y).

$$(2, -1)$$

Solving Systems of Linear Equations by Elimination

Recall that the addition and subtraction **properties of equality** state that two expressions in an equation will remain equal as long as the same operations are performed on both sides.

Addition	**Subtraction**
If $a = b$, then $a + c = b + c$.	If $a = b$, then $a - c = b - c$.

Using properties of equality

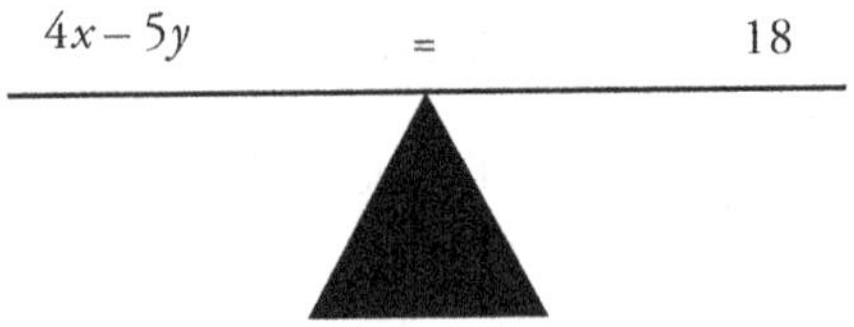

If $3x + 5y = -11$, then you could add $3x + 5y$ to one side of the equation and -11 to the other side of the equation, and both sides will remain equal.

$$\begin{array}{lcr} 4x - 5y & = & 18 \\ +(3x + 5y) & & +(-11) \\ \hline 7x & = & 7 \\ \hline \end{array}$$

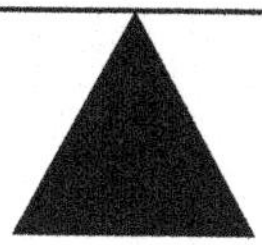

You can combine like terms and rewrite both equations as one equation:

$$7x = 7$$

Notice that the y-variable is eliminated, allowing you to solve for x.

$$\begin{aligned} 7x &= 7 \\ \frac{7x}{7} &= \frac{7}{7} \\ x &= 1 \end{aligned}$$

Once you solve for one variable, you can substitute that value into either of the original equations and solve for the other variable.

Let's substitute 1 for x into either of the original equations and solve for y.

$$\begin{array}{rcl} 4(1) - 5y = 18 & \qquad & 3(1) + 5y = -11 \\ 4 - 5y = 18 & & 3 + 5y = -11 \\ -4 \qquad\qquad -4 & & -3 \qquad\qquad -3 \\ -5y = 14 & & 5y = -14 \\ \frac{-5y}{-5} = \frac{14}{-5} & & \frac{5y}{5} = \frac{-14}{5} \\ y = -2.8 & & y = -2.8 \end{array}$$

The solution to this system of linear equations is (1, –2.8).

Example 7:

Solve by elimination.

$$\begin{cases} 7x - 4y = 13 \\ 2x + 4y = 5 \end{cases}$$

Solution:

Step 1. Line up like terms and combine into one equation by adding.

$$\begin{array}{r} 7x - 4y = 13 \\ +2x + 4y = \ \ 5 \\ \hline 9x \qquad = 18 \end{array}$$

Step 2. Solve for the variable.

$$\begin{array}{rl} \dfrac{9x}{9} = \dfrac{18}{9} & \text{Divide both sides by 9} \\ x = 2 & \text{Simplify} \end{array}$$

Step 3. Substitute the value of *x* into either equation and solve for *y*.

$$\begin{array}{rl} 2(2) + 4y = 5 & \text{Substitute 2 for } x \text{ and solve for } y \\ 4 + 4y = 5 & \text{Simplify (multiplication)} \\ -4 \qquad -4 & \text{Subtract 4 from both sides} \\ 4y = 1 & \text{Simplify} \\ \dfrac{4y}{4} = \dfrac{1}{4} & \text{Divide both sides by 4} \\ y = 0.25 & \text{Simplify} \end{array}$$

Step 4. Check by substituting both values into the other equation.

$$7(2) - 4(0.25) = 13 \quad \text{Substitute 2 for } x \text{ and 0.25 for } y$$
$$14 - 1 = 13 \quad \text{Simplify (multiplication)}$$
$$13 = 13 \quad \checkmark$$

Step 5. Write the solution as an ordered pair (x, y).

$$(2, 0.25)$$

Using multiplication property of equality

In Example 7, the y-terms $4y$ and $-4y$ eliminate each other since their coefficients are opposites.

If the coefficients of one variable are not opposites of each other, you must use the **multiplication property of equality** before you can use elimination.

Example 8:

Solve by elimination.

$$\begin{cases} 5x + 2y = 10 \\ 5x + 5y = 25 \end{cases}$$

Solution:

Step 1. Multiply the first equation by -1 to get opposite coefficients.

$$-1(5x + 2y) = -1(10)$$
$$-5x - 2y = -10$$

Step 2. Line up like terms and combine into one equation by adding.

$$\begin{array}{r} 5x + 5y = 25 \\ +(-5x - 2y = -10) \\ \hline 3y = 15 \end{array}$$

Step 3. Solve for the variable.

$$\frac{3y}{3} = \frac{15}{3} \quad \text{Divide both sides by 3}$$
$$y = 5 \quad \text{Simplify}$$

Step 4. Substitute the value of y into either equation and solve for x.

$$5x + 5(5) = 25 \quad \text{Substitute 5 for } y \text{ and solve for } x$$
$$5x + 25 = 25 \quad \text{Simplify (multiplication)}$$
$$-25 \quad -25 \quad \text{Subtract 25 from both sides}$$
$$5x = 0 \quad \text{Simplify}$$
$$\frac{5x}{5} = \frac{0}{5} \quad \text{Divide both sides by 5}$$
$$x = 0 \quad \text{Simplify}$$

Step 5. Check by substituting both values into the other equation.

$$5(0) + 2(5) = 10 \quad \text{Substitute 0 for } x \text{ and 5 for } y$$
$$0 + 10 = 10 \quad \text{Simplify (multiplication)}$$
$$10 = 10 \quad \checkmark$$

Step 6. Write the solution as an ordered pair (x, y).

$$(0, 5)$$

You can also multiply both equations to make them have opposite coefficients.

Example 9:

Solve by elimination.

$$\begin{cases} 3x+5y=17 \\ 4x-7y=9 \end{cases}$$

Solution:

Step 1. Multiply both equations to get opposite coefficients.

$7(3x+5y)=7(17)$ Multiply first equation by 7

$5(4x-7y)=5(9)$ Multiply second equation by 5

Step 2. Line up like terms and combine into one equation by adding.

$$\begin{array}{r} 21x+35y=119 \\ +(20x-35y=\ \ 45) \\ \hline 41x \qquad\quad =164 \end{array}$$

Step 3. Solve for the variable.

$\dfrac{41x}{41}=\dfrac{164}{41}$ Divide both sides by 41

$x=4$ Simplify

Step 4. Substitute the value of x into either equation and solve for y.

$4(4)-7y=9$ Substitute 4 for x and solve for y

$16-7y=9$ Simplify (multiplication)

$-16 \qquad -16$ Subtract 16 from both sides

$-7y=-7$ Simplify

$\dfrac{-7y}{-7}=\dfrac{-7}{-7}$ Divide both sides by -7

$y=1$ Simplify

Step 5. Check by substituting both values into the other equation.

$$\begin{aligned} 3(4)+5(1)&=17 && \text{Substitute 4 for } x \text{ and 1 for } y \\ 12+5&=17 && \text{Simplify (multiplication)} \\ 17&=17 && \checkmark \end{aligned}$$

Step 6. Write the solution as an ordered pair (x, y).

$$(4, 1)$$

Types of Solutions to Systems of Linear Equations

Recall that the solution to a system of linear equations is an ordered pair (x, y) at which all the lines intersect.

All of the examples we've seen thus far have had one solution. Now let's look at other types of solutions to systems of linear equations.

Intersecting lines | **Parallel lines** | **Coinciding lines**

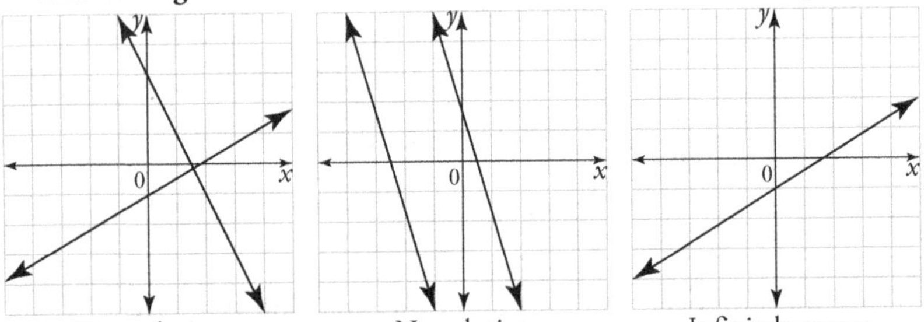

One solution | No solution | Infinitely many solutions

No solution

Example 10:

Solve by graphing.

$$\begin{cases} y=-2x+5 \\ 6x+3y=-12 \end{cases}$$

Solution:

Rewrite the second equation in slope-intercept form, $y = mx + b$, to find the y-intercept and the slope.

$$6x+3y=-12$$

$$-6x \qquad -6x \quad \text{Subtract } 6x \text{ from both sides}$$

$$3y=-6x-12 \quad \text{Simplify}$$

$$\frac{3y}{3}=-\frac{6x}{3}-\frac{12}{3} \quad \text{Divide both sides by 3}$$

$$y=-2x-4$$

$$\begin{cases} y=-2x+5 & y\text{-intercept of } 5, \text{ slope of } -2 \\ y=-2x-4 & y\text{-intercept of } -4, \text{ slope of } -2 \end{cases}$$

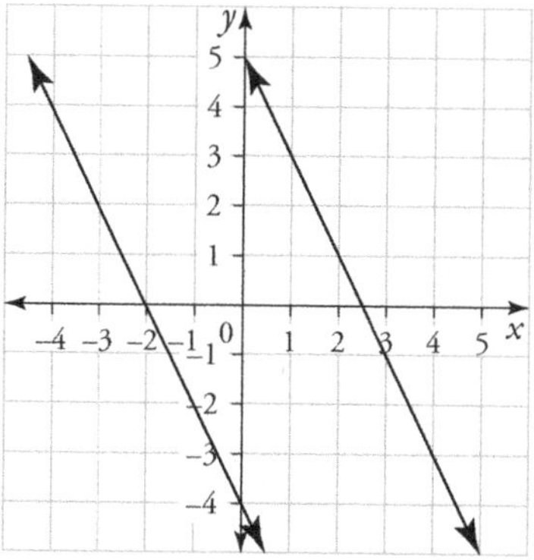

Notice that the lines are parallel and do NOT intersect. Therefore, this system of linear equations has **no solution.**

Infinitely many solutions

Example 11:

Solve by graphing.

$$\begin{cases} y=\frac{4}{7}x+2 \\ -4x+7y=14 \end{cases}$$

Solution:

Rewrite the second equation in slope-intercept form, $y = mx + b$, to find the y-intercept and the slope.

$$-4x + 7y = 14$$

$$+4x \qquad\qquad +4x \quad \text{Add } 4x \text{ to both sides}$$

$$7y = 4x + 14 \quad \text{Simplify}$$

$$\frac{7y}{7} = \frac{4x}{7} + \frac{14}{7} \quad \text{Divide both sides by 7}$$

$$y = \frac{4}{7}x + 2$$

$$\begin{cases} y = \frac{4}{7}x + 2 & y\text{-intercept of 2, slope of } \frac{4}{7} \\ y = \frac{4}{7}x + 2 & y\text{-intercept of 2, slope of } \frac{4}{7} \end{cases}$$

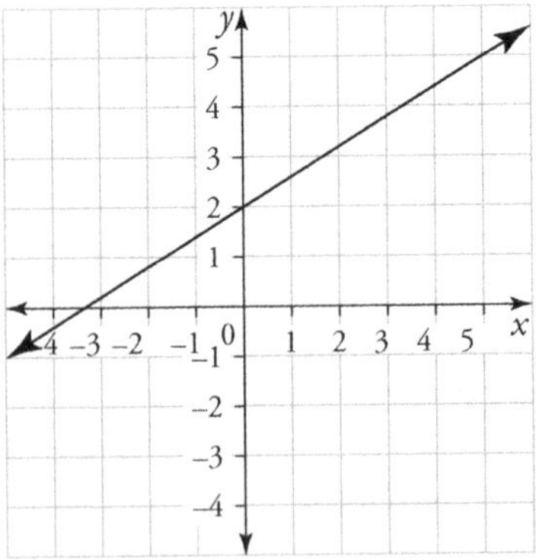

Notice that the two lines are coinciding. Therefore, this system of equations has **infinitely many solutions.**

Linear Inequalities

Recall that an **inequality** is a mathematical statement that two expressions are *not* equal.

$$a < b \qquad a > b \qquad a \leq b \qquad a \geq b$$

A linear inequality splits a number line into different regions and has one or more points that serve as boundaries. The **solutions of an inequality** are any values on a number line that make the inequality true.

$x > 7$

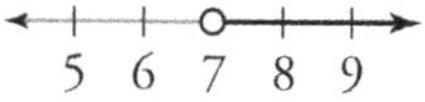

A **linear inequality** splits a coordinate grid into different regions with a line that serves as a boundary. The **solutions to a linear inequality** are any ordered pairs that make the inequality true.

Graphs of linear inequalities

Graph of $y = 2x + 1$

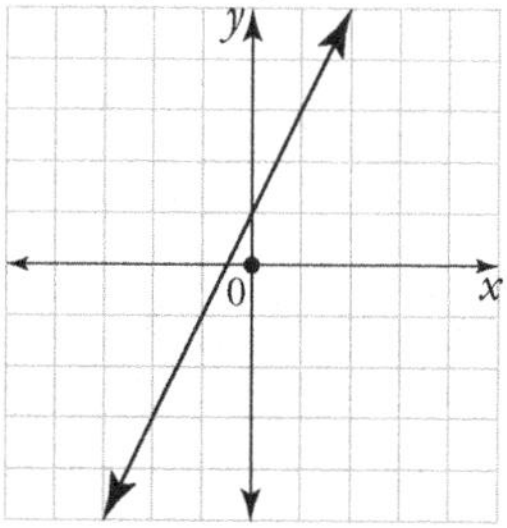

y-intercept of 1; slope of 2

Graph of $y > 2x + 1$

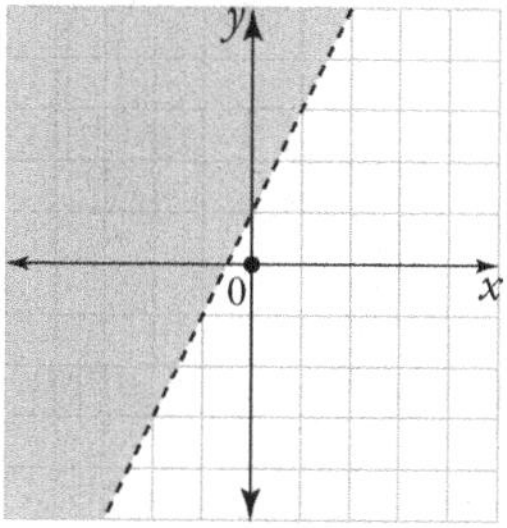

< or > means ordered pairs on line NOT included; shown by dashed line

> greater than symbol; shading *above* the line

Graph of $y \leq 2x + 1$

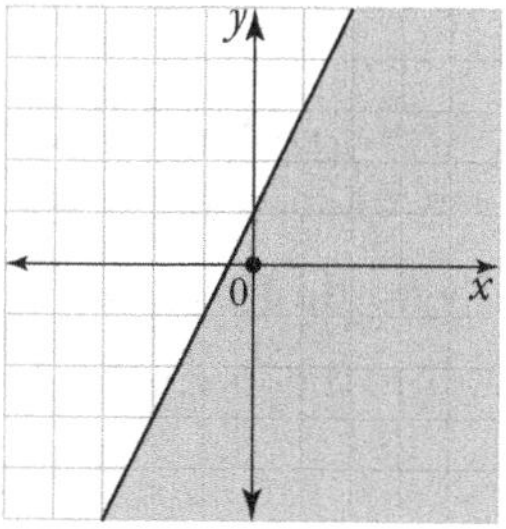

≤ or ≥ means ordered pairs on line ARE included; shown by solid line

≤ less than or equal to symbol; shading *below* the line

Example 12:

Graph $y < -x + 2$.

Solution:

Step 1. Graph points on line.

y-intercept of 2; slope of -1

Step 2. Select dashed or solid line.

< or > means ordered pairs on line NOT included; shown by dashed line

Step 3. Determine if it should be shaded above or below the line.

< less than symbol; shade *below* the line

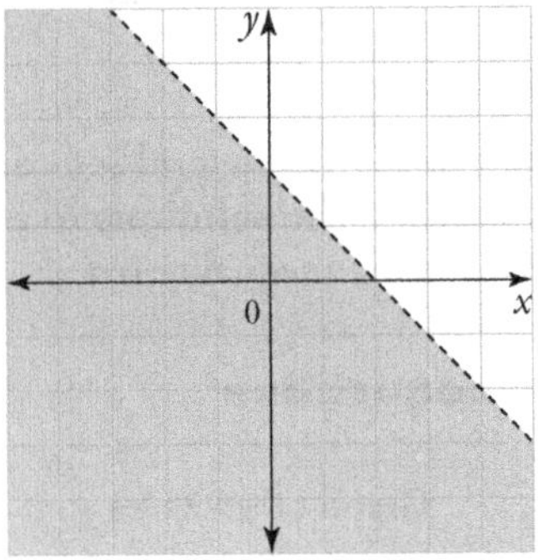

When graphing a linear inequality, the origin, (0, 0), can often be an easy way of checking if you have shaded correctly.

Example 13:

Write the inequality that represents this graph.

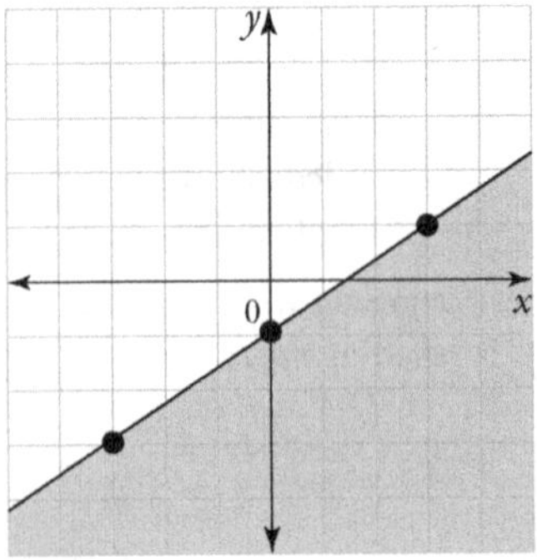

Solution:

Step 1. Identify points on line.

y-intercept of -1; slope of $\frac{2}{3}$

Step 2. Identify dashed or solid line.

$\leq$ or $\geq$ means ordered pairs on line ARE included; shown by *solid* line

Step 3. Determine if it should be shaded above or below the line.

$\leq$ less than or equal to symbol; shade *below* the line

The graph represents the inequality $y \leq \frac{2}{3}x - 1$.

Using a graph to determine ordered pair solutions

Recall that solutions of linear inequalities are ordered pairs that make the inequality true. You can use the graph of an inequality to determine which ordered pairs are solutions.

Example 14:

Which of the following are solutions of $y \leq \frac{2}{3}x - 1$?

A. (0, 0)

B. (4, –1)

C. (–3, –3)

Solution:

A. (0, 0)	**B.** (4, – 1)	**C.** (–3, –3)
Not in the shaded region.	In the shaded region.	Points on a solid line ARE included.
Not a solution.	Yes, it is a solution.	Yes, it is a solution.

The ordered pairs in items B and C are solutions to the equation $y \leq \frac{2}{3}x - 1$.

Substituting ordered pairs

You can also substitute ordered pairs (x, y) into a linear inequality to determine if they make the inequality true.

Example 15:

Which of the following are solutions of $y > 4x + 6$?

A. (0, 0)

B. (–1, 2)

C. (3, 20)

Solution:

A. (0, 0)	B. (–1, 2)	C. (3, 20)
Substitute 0 for x and 0 for y	Substitute – 1 for x and 2 for y	Substitute 3 for x and 20 for y
$0 > 4(0)+6$	$2 > 4(-1)+6$	$20 > 4(3)+6$
$0 > 0+6$	$2 > -4+6$	$20 > 12+6$
$0 > 6$	$2 > 2$	$20 > 18$
False, 0 is NOT greater than 6.	False, 2 is NOT greater than 2.	True, 20 is greater than 18.
Not a solution.	Not a solution.	Yes, it is a solution.

The ordered pair in item C is a solution to the equation $y > 4x + 6$.

Systems of Linear Inequalities

When you graph two or more linear inequalities on the same coordinate grid, you create a **system of linear inequalities.** The **solutions to a system of linear inequalities** are any ordered pairs that make all inequalities true.

Take a look at the graph for the following system of linear inequalities:

$$\begin{cases} y < \frac{3}{2}x \\ y < -x - 3 \end{cases}$$

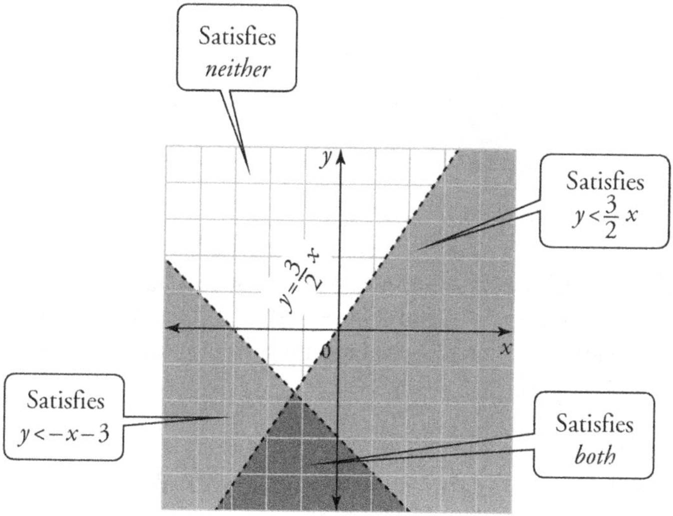

All points in the region where shading overlaps make both inequalities true.

Example 16:

$$\begin{cases} y \geq 2x + 4 \\ y < -3x \end{cases}$$

A. Graph the system of linear inequalities.

B. Identify if the ordered pairs (–1, 0) and (–3, 2) are solutions to the system.

Solution:

A.

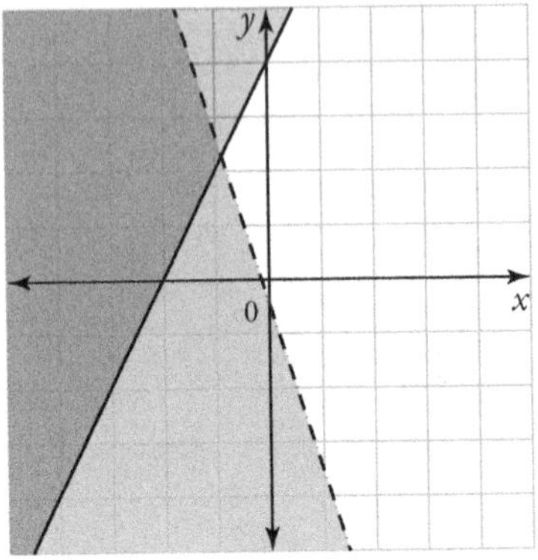

B. $(-1,0)$

Substitute -1 for x and 0 for y in the first inequality

$0 \geq 2(-1)+4$

$0 \geq -2+4$

$0 \geq 2$

False, 0 is NOT greater than or equal to 2.

Not a solution.

$(-3,2)$

Substitute -3 for x and 2 for y in the first inequality

$2 \geq 2(-3)+4$

$2 \geq -6+4$

$2 \geq -2$

True, 2 is greater than or equal to -2.

Substitute -3 for x and 2 for y in the second inequality

$2 < -3(-3)$

$2 < 9$

True, 2 is less than 9.

(–3, 2) is a solution to the system because it makes both inequalities true.

TECH TIPS: Solving Systems of Linear Equations and Graphing Linear Inequalities on a Graphing Calculator

Solving systems of linear equations

Recall that the solution to a system of equations is any point that makes all equations in the system true. When you graph both equations on the same coordinate grid, the solution is the **point of intersection** between those lines.

Example:

Solve:

$$\begin{cases} y = \dfrac{4}{5}x + 1 \\ y = -2x + 9 \end{cases}$$

Solution:

1. Press [Y =] and enter the first equation into Y_1=.

   Enter the second equation into Y_2=.

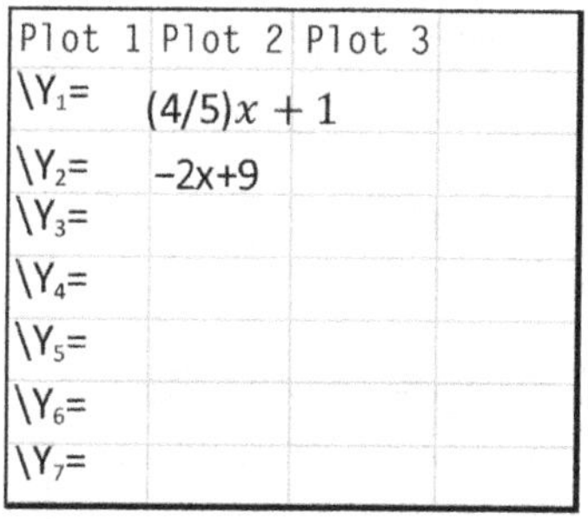

2. Press [GRAPH].

   Notice that the graphs intersect at a point in Quadrant I.

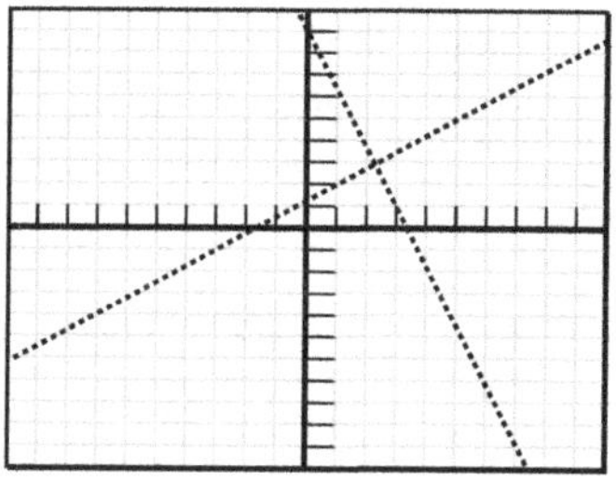

3. Press [2nd] [TRACE].

4. While in CALC mode, press **5** to calculate the point of intersection.

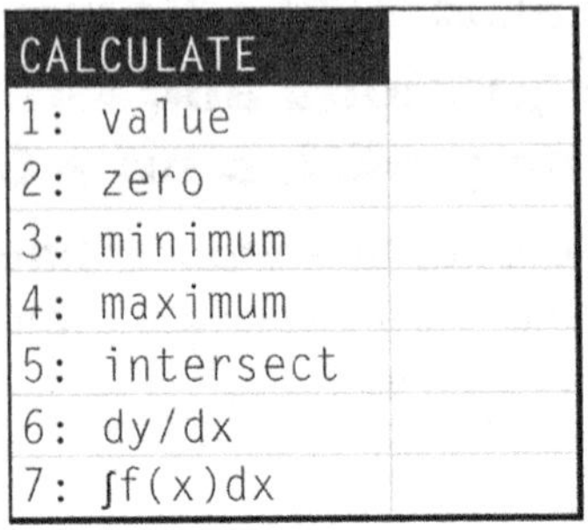

5. Press **2nd**.
6. The screen will prompt you with three questions:

   First curve?
   Second curve?
   Guess?

   Press **ENTER** three times.

   The point of intersection will appear at the bottom of the screen.

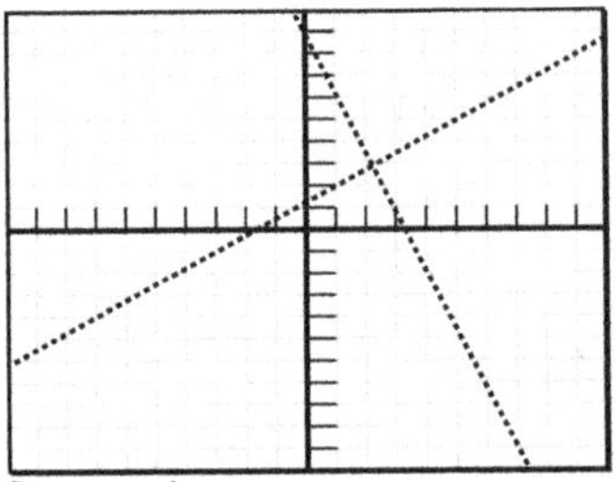

Intersection
X = 2.8571429   Y = 3.287143

Answer: The point of intersection, rounded to the nearest one hundredth, is (2.86, 3.29).

Graphing inequalities

Recall that linear inequalities are represented by shaded regions on a coordinate grid that have lines as boundaries.

Example:

Graph the solutions to the inequality $y > 3x - 6$.

Solution:

1. Press **Y=** and enter the expression into Y_1=.

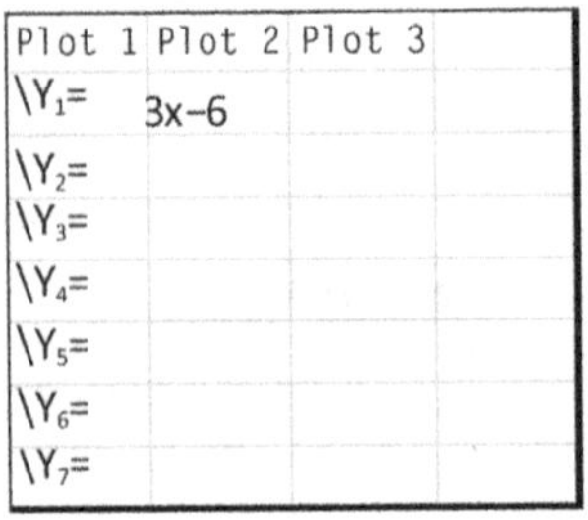

2. The inequality reads "*y* is *greater than*" so you will shade *above* the line.

   Press the left arrow key once to highlight the slanted line to the left of Y_2 and change the line settings.

   Press **ENTER** *two* times to change the default "thin line" setting to the "shade above" setting.

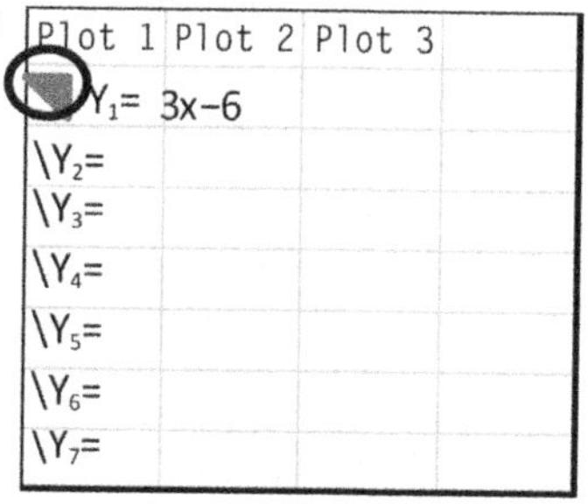

3. Press **GRAPH**.

   Notice that the line will graph first, followed by the shading above the line.

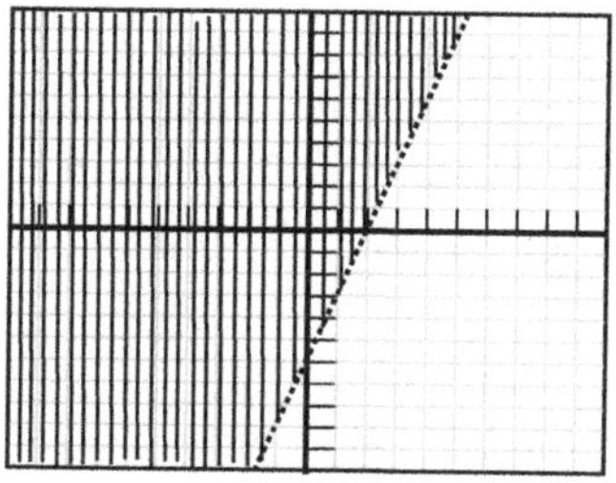

Answer: The solutions to the inequality are represented by the shaded region.

Chapter Check-Out

Questions

For questions 1–3, solve by graphing.

1. $\begin{cases} y = -x + 1 \\ y = \dfrac{1}{3}x - 3 \end{cases}$

2. $\begin{cases} y = 7x - 4 \\ -14x + 2y = -8 \end{cases}$

3. $\begin{cases} y = -\dfrac{1}{2}x + 11 \\ x + 2y = 6 \end{cases}$

4. Solve by substitution:

$$\begin{cases} y = 5x - 9 \\ 5x - 5y = -5 \end{cases}$$

5. Solve by elimination:

$$\begin{cases} x - y = 8 \\ 2x + y = 4 \end{cases}$$

6. Adrian has a total of 60 coins in his coin jar. The coins, made up of quarters and dimes, add up to $12. Write and solve a system of equations to determine how many quarters, x, and how many dimes, y, Adrian has in his coin jar.

7. Which of the following is a solution to the inequality $y > \frac{1}{3}x - 5$?

A. (8, –5)
B. (0, –5)
C. (–1, –3)

8. Which of the following is a solution to the following system of inequalities?

$$\begin{cases} y < -3x + 9 \\ y \geq 5x - 2 \end{cases}$$

A. (2, –4)
B. (0, 0)
C. (1, 9)

Answers

1. (3, –2)
2. infinitely many solutions
3. no solution
4. (2.5, 3.5)
5. (4, –4)
6. $\begin{cases} x + y = 60 \\ 0.25x + 0.10y = 12 \end{cases}$

   (40, 20)
7. C
8. B

Chapter 7
EXPONENTS

Chapter Check-In

- ❑ Properties of exponents
- ❑ Zero and negative exponents
- ❑ Multiplication properties of exponents
- ❑ Division properties of exponents
- ❑ Exponential functions

In science, sometimes you work with very large or very small numbers. We can use exponents and scientific notation to represent those very large or very small numbers. In order to solve problems involving scientific notation, let's explore the properties of exponents.

Properties of Exponents

You can write the multiplication of the same number using **exponential form.**

$$2 \cdot 2 \cdot 2 \cdot 2 \cdot 2 = 2^5 \qquad 5 \cdot 5 = 5^2$$

In exponential form, the number being multiplied is called the **base.** The **exponent** represents the number of times it is being multiplied by itself.

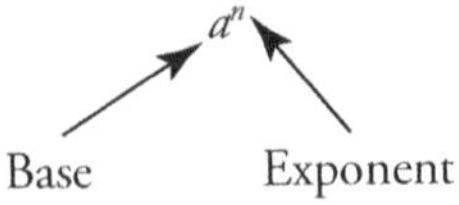

You read numbers in exponential form as "*a* to the *n*th power."

3^4 read "three to the fourth power"

6^5 read "six to the fifth power"

Exponents of 2 and 3 have special names.

7^2	9^3
read "seven squared"	read "nine cubed"

You can evaluate a number in exponential form by multiplying the base, *a*, by itself *n* number of times.

Exponential Form	Expanded Form	Standard Form
4^5	$4 \cdot 4 \cdot 4 \cdot 4 \cdot 4$	1,024
8^3	$8 \cdot 8 \cdot 8$	512

Notice that as the exponent increases by increments of one, the number is simply being multiplied by the base an additional time.

As you decrease the exponent by increments of one, the value of the number is being divided by the base.

Increasing and Decreasing Exponents

	Exponential Form	Expanded Form	Standard Form	
	2^4	$2 \cdot 2 \cdot 2 \cdot 2$	16	
×2				÷2
	2^3	$2 \cdot 2 \cdot 2$	8	
×2				÷2
	2^2	$2 \cdot 2$	4	
×2				÷2
	2^1	2	2	

Zero and Negative Exponents

Continue to decrease the exponent by increments of one and divide by the base to observe what happens when you reach zero and negative exponents.

Zero and Negative Exponents

Exponential Form	Expanded Form	Standard Form
2^1	2	2
2^0	$\frac{2}{2}$	1
2^{-1}	$\frac{1}{2^1}$	$\frac{1}{2}$
2^{-2}	$\frac{1}{2 \cdot 2} = \frac{1}{2^2}$	$\frac{1}{4}$

÷2
÷2
÷2

Zero Exponent:

Any nonzero number, a, raised to the power of 0 is equal to 1.

$$a^0 = 1, \text{ when } a \neq 0$$

Negative Exponents:

Any nonzero number, a, raised to a negative exponent is equal to 1 divided by that number raised to that exponent's opposite (positive) value.

$$a^{-n} = \frac{1}{a^n}, \text{ when } a \neq 0$$

Example 1:

Simplify each of the following expressions.

A. $(-4)^0$

B. 7^{-2}

Solution:

A. Any nonzero number raised to the power of 0 is equal to 1.
$(-4)^0 = 1$

B. $7^{-2} = \frac{1}{7^2}$ Use property of negative exponents

$= \frac{1}{49}$ Simplify

Simplifying expressions with multiple terms and exponents

To simplify expressions with multiple terms and exponents, write the expression as a product of separate terms and simplify using positive exponents only.

Example 2:

Simplify the expressions.

A. $6y^{-3}$

B. $\frac{g^{-5}}{d^4}$

C. $9p^{-7}r^8$

D. $\frac{11}{c^{-2}}$

Solution:

A. $6y^{-3}$

$6 \cdot y^{-3}$	Rewrite as a product of two terms
$6 \cdot \frac{1}{y^3}$	Use property of negative exponents
$\frac{6}{y^3}$	Simplify

B. $\frac{g^{-5}}{d^4}$

$\frac{g^{-5}}{1} \cdot \frac{1}{d^4}$	Rewrite as a product of separate terms
$\frac{1}{g^5} \cdot \frac{1}{d^4}$	Use property of negative exponents
$\frac{1}{d^4 g^5}$	Simplify

C. $9p^{-7}r^{8}$

$9 \cdot p^{-7} \cdot r^{8}$ Rewrite as a product of separate terms

$\frac{9}{1} \cdot \frac{1}{p^{7}} \cdot \frac{r^{8}}{1}$ Use property of negative exponents

$\frac{9r^{8}}{p^{7}}$ Simplify

D. $\frac{11}{c^{-2}}$

$\frac{11}{1} \cdot \frac{1}{c^{-2}}$ Rewrite as the product of separate terms

$\frac{11}{1} \cdot \frac{1}{\left(\frac{1}{c^{2}}\right)}$ Use property of negative exponents

$\frac{11}{1} \cdot \frac{c^{2}}{1}$ To divide by a fraction, multiply by the reciprocal

$11c^{2}$ Simplify

Multiplication Properties of Exponents

When multiplying numbers in exponential form, you can simplify them as one power using properties of exponents if their bases are the same.

Multiplying Numbers in Exponential Form		
Exponential Form	***Expanded Form***	***Simplified***
$2^2 \cdot 2^3$	$(2 \cdot 2) \cdot (2 \cdot 2 \cdot 2)$	2^5
$8^4 \cdot 8^3$	$(8 \cdot 8 \cdot 8 \cdot 8) \cdot (8 \cdot 8 \cdot 8)$	8^7
$x \cdot x^5$	$(x) \cdot (x \cdot x \cdot x \cdot x \cdot x)$	x^6
$a^m \cdot a^n$		a^{m+n}

Notice that when multiplying two numbers in exponential form with the *same base*, you can simplify as one power by adding the exponents.

Multiplication Property of Exponents:

For any nonzero base, *a*, with integer exponents *m* and *n*,

$$a^m \cdot a^n = a^{m+n}$$

Example 3:

Simplify each expression by writing it with only one base.

A. $13^9 \cdot 13^8$

B. $7^2 \cdot 7^6 \cdot 7^{-8}$

Solution:

A. $13^9 \cdot 13^8$

13^{9+8}	Use multiplication property of exponents
13^{17}	Simplify addition

B. $7^2 \cdot 7^6 \cdot 7^{-8}$

$7^{2+6+(-8)}$	Use multiplication property of exponents
7^0	Simplify addition
1	Use zero property of exponents

Simplifying the product of expressions with different bases

When simplifying the product of expressions with different bases, combine the constants and powers with the same base.

Example 4:

Simplify each expression.

A. $3p^5 \cdot 5p^6$

B. $(4c) \cdot (8b^3) \cdot (2c^9)$

Solution:

A. $3p^5 \cdot 5p^6$

$(3 \cdot 5) \cdot (p^5 \cdot p^6)$	Group constants and powers with the same base
$15 \cdot p^{5+6}$	Use multiplication property of exponents
$15p^{11}$	Simplify addition

B. $(4c) \cdot (8b^3) \cdot (2c^9)$

$(4 \cdot 8 \cdot 2) \cdot (b^3) \cdot (c \cdot c^9)$	Group constants and powers with the same base
$64 \cdot b^3 \cdot c^{1+9}$	Use multiplication property of exponents
$64b^3c^{10}$	Simplify addition

Raising a power to a power

Sometimes you can view an exponential function as the base of another exponential function. We refer to this as **raising a power to a power.**

Raising a Power to a Power

Exponential Form	*Expanded Form*	*Simplified*
$(2^2)^3$	$(2 \cdot 2) \cdot (2 \cdot 2) \cdot (2 \cdot 2)$	2^6
$(8^4)^2$	$(8 \cdot 8 \cdot 8 \cdot 8) \cdot (8 \cdot 8 \cdot 8 \cdot 8)$	8^8
$(x^2)^5$	$(x \cdot x) \cdot (x \cdot x) \cdot (x \cdot x) \cdot (x \cdot x) \cdot (x \cdot x)$	x^{10}
$(a^m)^n$		$a^{m \cdot n}$

Raising a Power to a Power:

For any nonzero base, *a*, with integer exponents *m* and *n*,

$$(a^m)^n = a^{m \cdot n}$$

Example 5:

Simplify each expression.

A. $(b^5)^6$

B. $(k^3)^{-4}$

Solution:

A. $(b^5)^6$

$b^{5\cdot 6}$	Use property of raising a power to a power
b^{30}	Simplify (multiplication)

B. $(k^3)^{-4}$

$k^{3\cdot(-4)}$	Use property of raising a power to a power
k^{-12}	Simplify (multiplication)
$\frac{1}{k^{12}}$	Use property of negative exponents

Division Properties of Exponents

When dividing numbers in exponential form, you can simplify them as one power using properties of exponents if their bases are the same.

Dividing Numbers in Exponential Form		
Exponential Form	***Expanded Form***	***Simplified***
$\frac{2^5}{2^3}$	$\frac{\cancel{2}\cdot\cancel{2}\cdot\cancel{2}\cdot 2\cdot 2}{\cancel{2}\cdot\cancel{2}\cdot\cancel{2}}$	2^2
$\frac{8^4}{8^3}$	$\frac{\cancel{8}\cdot\cancel{8}\cdot\cancel{8}\cdot 8}{\cancel{8}\cdot\cancel{8}\cdot\cancel{8}}$	8
$\frac{x^5}{x}$	$\frac{\cancel{x}\cdot x\cdot x\cdot x\cdot x}{\cancel{x}}$	x^4
$\frac{a^m}{a^n}$		a^{m-n}

Notice that when dividing two numbers in exponential form with the same base, you can simplify as one power by subtracting the exponents.

Division Property of Exponents:

For any nonzero base, a, with integer exponents m and n,

$$\frac{a^m}{a^n} = a^{m-n}$$

Example 6:

Simplify each expression.

A. $\frac{z^6}{z^{13}}$

B. $\frac{d^5 g^2}{d^3 g^6}$

Solution:

A. $\frac{z^6}{z^{13}}$

z^{6-13}	Use division property of exponents
z^{-7}	Simplify (subtraction)
$\frac{1}{z^7}$	Use property of negative exponents

B. $\frac{d^5 g^2}{d^3 g^6}$

$(d^{5-3}) \cdot (g^{2-6})$	Group powers with the same base; use division property of exponents
$d^2 g^{-4}$	Simplify (subtraction)
$d^2 \cdot \frac{1}{g^4}$	Use property of negative exponents
$\frac{d^2}{g^4}$	Simplify

Exponential Functions

A function with an independent variable as an exponent is called an **exponential function.**

Exponential Function:
A function in the form $y = a \cdot b^x$, where $a \neq 0$, $b > 0$ and $b \neq 1$, and x is a real number.

Example 7:

Evaluate each exponential function for the domain {–2, –1, 0, 1, 2}. Make a table of values and graph each function.

A. $y = 2^x$

B. $y = 5 \cdot 3^x$

Solution:

A. $y = 2^x$

x	2^x	y
–2	2^{-2}	$\frac{1}{4}$
–1	2^{-1}	$\frac{1}{2}$
0	2^0	1
1	2^1	2
2	2^2	4

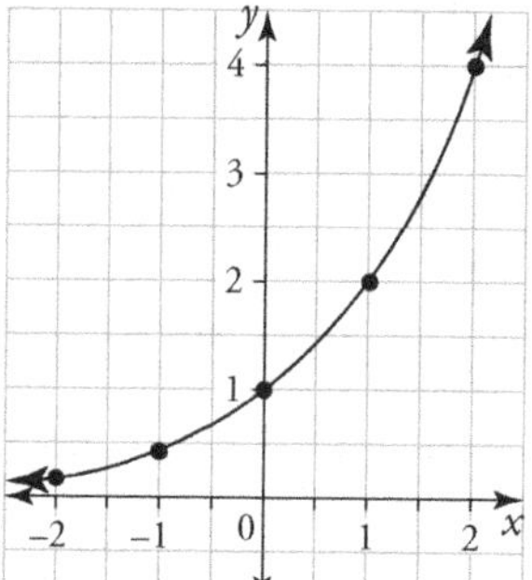

B. $y = 5 \cdot 3^x$

x	$5 \cdot 3^x$	y
–2	$5 \cdot 3^{-2} = 5 \cdot \frac{1}{9}$	$\frac{5}{9}$
–1	$5 \cdot 3^{-1} = 5 \cdot \frac{1}{3}$	$\frac{5}{3}$
0	$5 \cdot 3^0 = 5 \cdot 1$	5
1	$5 \cdot 3^1 = 5 \cdot 3$	15
2	$5 \cdot 3^2 = 5 \cdot 9$	45

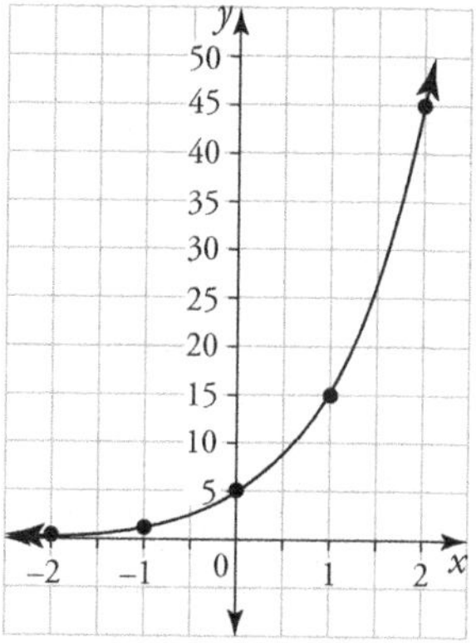

If you connect the points and extend the graph in both directions, the domain of an exponential function is all real numbers, or

$$\text{Domain: } (-\infty, \infty)$$

The range is $y > 0$, or

$$\text{Range: } (0, \infty)$$

Notice that as the value of the exponent, x, decreases, the value of the dependent variable, y, gets closer and closer to the horizontal line $y = 0$. The value that a function approaches but doesn't reach is called an **asymptote.**

Exponential growth

Exponential functions can also be used to model populations over time.

> **Exponential Growth:**
>
> A function in the form $y = a \cdot b^x$
>
> (a: Initial value; b: Growth rate)
>
> where $a \neq 0$, and $b > 0$ and $b \neq 1$.

Example 8:

It is estimated that mice can have a litter of up to six young every 30 days. The potential population of mice can be modeled by the following equation, where y represents the population size and x represents elapsed time in months.

$$y = 2 \cdot 6^x$$

(2: Initial value; 6: Growth rate)

Solution:

Find the potential population for the first 3 months. Make a table of values and graph the function.

x	$2 \cdot 6^x$	y
0	$2 \cdot 6^0 = 2 \cdot 1$	2
1	$2 \cdot 6^1 = 2 \cdot 6$	12
2	$2 \cdot 6^2 = 2 \cdot 36$	72
3	$2 \cdot 6^3 = 2 \cdot 216$	432

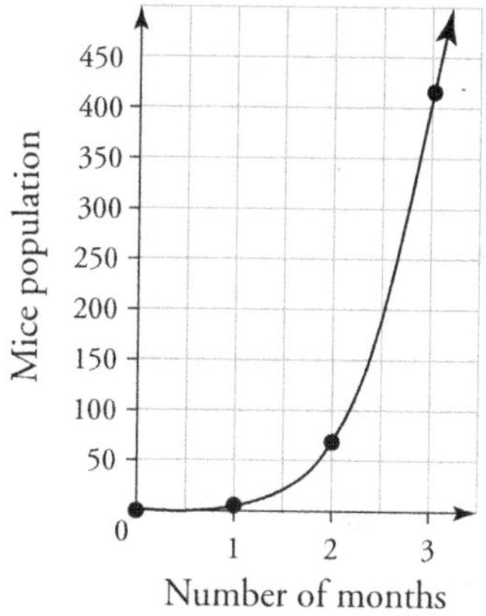

Notice that the initial value, a, is also the y-intercept (0, 2).

The domain of this function is Domain: $x \geq 0$ since time elapsed can only be described using positive integers.

The range of this function is Range: $y \geq 2$ since the initial population is 2 and could theoretically increase infinitely.

Exponential decay

When the base, b, is a value less than 1 but greater than 0, the value of the function decreases exponentially. This is known as **exponential decay.**

Exponential Decay:

A function in the form $y = a(1 - b)^x$

Initial value (arrow to a) Decay rate (arrow to $(1 - b)$)

where $a \neq 0$, and $0 < b < 1$.

Example 9:

The amount of time it takes your body to eliminate caffeine from your system is approximately 5 hours. On average, a cup of coffee has 95 mg of caffeine. The amount of caffeine that remains in your body, $f(x)$, after drinking one cup of coffee can be modeled by the function $f(x) = 95(0.5)^x$, where x represents the number of 5-hour periods that have elapsed.

Solution:

Make a table of values for the amount of caffeine remaining in your body the first 15 hours after drinking a cup of coffee. (***Hint:*** 15 hours = three 5-hour periods.) Use domain values {0, 1, 2, 3}.

x	$95(0.5)^x$	$f(x)$
0	$95(0.5)^0 = 95(1)$	95
1	$95(0.5)^1 = 95(0.5)$	47.5
2	$95(0.5)^2 = 95(0.25)$	23.75
3	$95(0.5)^3 = 95(0.125)$	11.875

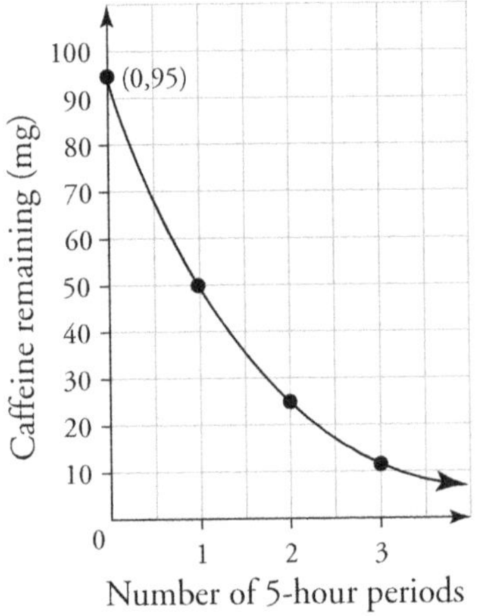

Notice that the initial value, a, is also the y-intercept (0, 95).

The domain of this function is Domain: $x \geq 0$ since time elapsed can only be described using positive integers.

The range of this function is Range: $0 < y \leq 95$ since the initial amount of caffeine is 95 mg and the function approaches 0.

Chapter Check-Out

Questions

For questions 1–2, simplify.

1. $\left(\frac{4}{5}\right)^0$
2. $5p^0$

For questions 3–4, evaluate the expression.

3. m^{-3}, for $m = 4$
4. $7b^{-5}$, for $b = 2$

For questions 5–10, simplify.

5. $x \cdot y^{-3} \cdot x^4$
6. $(k^7)^{-2}$
7. $(2w^8)^3$
8. $(g^9h^{-4})^6$
9. $\frac{7^5}{7^3}$
10. $\frac{q^5r^3}{(qr)^4}$

For questions 11–12, evaluate each function.

11. $f(x) = 9^x$, for $x = 3$
12. $d(g) = 6 \cdot 2^g$, for $g = 4$

For questions 13–15, find the range of the function for the given domain.

13. $h(w) = -2 \cdot 7^w$, for domain of $\{-2, -1, 0, 1, 2\}$
14. A bacteria culture starts with a population of 1,000 bacteria and triples every hour. Write an exponential growth function that represents the relationship between the number of hours, h, and the population of bacteria, p. Find the population of bacteria after 5 hours.
15. Rodrigo purchased a new car for $23,000. The value of the car depreciates at a rate of 15% per year. Write an exponential decay function that represents the relationship between the number of years, y, and the value of the car, v. Find the value of Rodrigo's car after 6 years.

Answers

1. 1
2. 5
3. $\frac{1}{64}$
4. $\frac{7}{32}$
5. $\frac{x^5}{y^3}$
6. $\frac{1}{k^{14}}$
7. $8w^{24}$
8. $\frac{g^{54}}{h^{24}}$
9. 49
10. $\frac{q}{r}$
11. 729
12. 96
13. $\text{Range}: \left\{\frac{-2}{49}, \frac{-2}{7}, -2, -14, -98\right\}$
14. $p = 1{,}000(3)^h$; 243,000 bacteria
15. $v = 23{,}000(0.85)^y$; \$8,674.44

Chapter 8
POLYNOMIALS

Chapter Check-In

- ❑ Adding and subtracting polynomials
- ❑ Multiplying polynomials
- ❑ Factoring polynomials
- ❑ Factoring trinomials

A **monomial** is a constant, a variable, or the product of a number and one or more variables.

$$14 \qquad x \qquad -8xy \qquad 7x^2y^3$$

The degree of a monomial is the sum of the exponents of the variables. If the monomial is a nonzero constant, the degree is 0.

Example 1:

Find the degree of each monomial.

A. 9

B. $-7x$

C. $6p^3q^5$

Solution:

A. The monomial is a nonzero constant, the degree is 0.

B. The implied exponent on the variable is 1, the degree is 1.

C. The exponents are 3 and 5, $3 + 5 = 8$; the degree is 8.

A **polynomial** is the sum or difference of two or more monomials.

To write a polynomial in **standard form,** combine like terms and arrange the monomials in order from the greatest to least degree.

The **degree of a polynomial** is also the degree of the monomial with the greatest degree.

When written in standard form, you can name a polynomial based on its degree and the number of terms it has.

Degree	**Name**
0	Constant
1	Linear
2	Quadratic
3	Cubic
4	Quartic
5	Quintic
n, used for 6 or more	*n*th degree

Number of Terms	**Name**
1	Monomial
2	Bi nomial
3	Trinomial
4 or more	Polynomial

Example 2:

Write each polynomial in standard form and name it based on its degree and its number of terms.

A. $2x + 5x^2 - 7$

B. $2x^3 - 8 + x^3$

C. $11x^5$

Solution:

A. $2x + 5x^2 - 7$

$5x^2 + 2x - 7$ Arrange from greatest to least degree

Highest degree: 2

Number of terms: 3

Quadratic trinomial

B. $2x^3 - 8 + x^3$

$3x^3 - 8$ Combine like terms

Highest degree: 3

Number of terms: 2

Cubic binomial

C. Highest degree: 5

Number of terms: 1

Quintic monomial

Adding and Subtracting Polynomials

To add or subtract polynomials, line up like terms and add or subtract using the coefficients of the variable terms.

Example 3:

Simplify the sum of polynomials.

A. $(3x^2 + 9x - 1) + (x^2 - 4x + 6)$

B. $(9z^2 + z - 4) + (7z^2 + 8)$

Solution:

A.

$3x^2 + 9x - 1$	
$+ x^2 - 4x + 6$	Line up like terms
$4x^2 + 5x + 5$	Simplify the sum of the coefficients

B.

$9z^2 + z - 4$	
$+ 7z^2 \quad + 8$	Line up like terms
$16z^2 + z + 4$	Simplify sum

To simplify the difference of polynomials, you can add the opposite of the polynomial being subtracted.

Example 4:

Simplify the difference of polynomials: $(4m^3 + 6m - 11) - (3m^3 - 2m^2 + 4)$.

Solution:

$4m^3 \quad + 6m - 11$	
$-(3m^3 - 2m^2 \quad + 4)$	Line up like terms

$4m^3 \quad + 6m - 11$	
$+(-3m^3 + 2m^2 \quad - 4)$	Rewrite as the sum of the opposite
$m^3 + 2m^2 + 6m - 15$	Simplify sum

Multiplying Polynomials

Using the distributive property

You can use the distributive property and the multiplication property of exponents to simplify the product of a monomial and a polynomial.

Example 5:

Simplify $5v(2v^2 - 3v + 6)$.

Solution:

$5v\,(2v^2 - 3v + 6)$

$(5v)\cdot(2v^2) + (5v)\cdot(-3v) + (5v)\cdot(6)$ Use distributive property

$10v^3 - 15v^2 + 30v$ Use multiplication property of exponents

You can also use the distributive property to multiply polynomials.

Example 6:

Simplify $(x + 3)(x + 4)$.

Solution:

$x(x + 4) + 3(x + 4)$ Distribute the x and distribute the 3

$(x^2 + 4x) + (3x + 12)$ Use multiplication property of exponents

$$\begin{array}{r} x^2 + 4x \\ +(\phantom{x^2+{}} 3x + 12) \\ \hline x^2 + 7x + 12 \end{array}$$

Line up like terms and add

Simplify sum and write in standard form

FOIL method

Another method for multiplying binomials is the **FOIL** method. FOIL is an acronym that stands for the order in which you multiply the terms of the two binomials: "First, Outer, Inner, Last."

Let's use FOIL to simplify$(x - 7)(x + 2)$.

$(x - 7)(x + 2)$ Multiply First terms $x \cdot x = x^2$

$(x - 7)(x + 2)$ Multiply Outer terms $x \cdot 2 = 2x$

$(x-7)(x+2)$	Multiply Inner terms	$-7 \cdot x = -7x$
$(x-7)(x+2)$	Multiply Last terms	$-7 \cdot 2 = -14$
$x^2 + 2x - 7x - 14$	Write all four products as one polynomial	
$x^2 - 5x - 14$	Combine like terms and simplify	

Example 7:

Simplify using FOIL: $(2x + 4)(3x - 5)$.

Solution:

$(2x+4)(3x-5)$	Multiply the First, Outer, Inner, Last terms
$2x(3x) + 2x(-5) + 4(3x) + 4(-5)$	
$6x^2 - 10x + 12x - 20$	Use multiplication property of exponents
$6x^2 + 2x - 20$	Combine like terms and write in standard form

Area models

Example 8:

Find the area of a rectangle with a length of $x - 8$ and a width of $x + 1$.

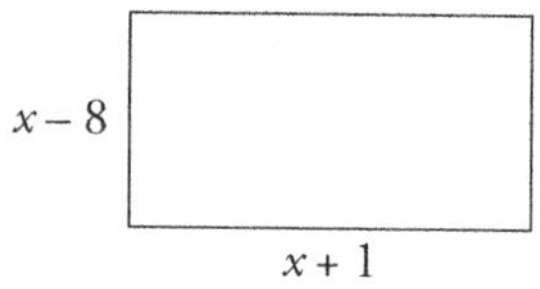

Solution:

$A = lw$	Use the formula for the area of a rectangle
$A = (x-8)(x+1)$	Substitute $x-8$ for l and $x+1$ for w
$A = x(x) + x(1) - 8(x) - 8(1)$	Simplify using FOIL
$A = x^2 + x - 8x - 8$	Use multiplication property of exponents
$A = x^2 - 7x - 8$	Combine like terms and write in standard form

Squares of binomials

To simplify the square of a binomial, rewrite it as the product of two binomials.

Example 9:

Find the area of a square with a side length of $x + 7$.

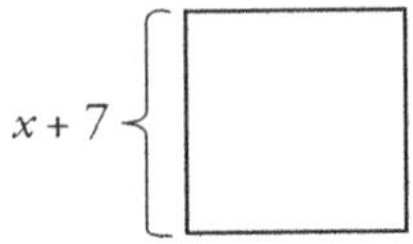

Solution:

$A = s^2$	Use the formula for the area of a square
$A = (x+7)^2$	Substitute $(x+7)$ for s
$A = (x+7)(x+7)$	Rewrite as the product of two binomials
$A = x(x+7)+7(x+7)$	First distribute x, then distribute 7
$A = x^2+7x+7x+49$	Use multiplication property of exponents
$A = x^2+14x+49$	Combine like terms and write in standard form

Notice that the square of a binomial is the sum of the first term squared, the second term times the first term doubled, and the second term squared. (***Note:*** The FOIL method could also be used to solve Example 9.)

> **Square of a Binomial:**
>
> $$(a + b)^2 = a^2 + 2ab + b^2$$

Let's look at one more special case of binomial multiplication.

Example 10:

Simplify $(x + 9)(x - 9)$.

Solution:

$x(x)+x(-9)+9(x)+9(-9)$	Multiply using FOIL method
$x^2-9x+9x-81$	Use multiplication property of exponents
x^2-81	Simplify sum and write in standard form

When you multiply the sum and difference of the same two terms, the product is called the **difference of squares.**

Notice that the two linear terms are opposites and their sum is zero, eliminating the linear term. The solution is made up of the difference of the first term squared and the second term squared.

Difference of Squares:

$$(a + b)(a - b) = a^2 - b^2$$

Factoring Polynomials

Factoring is the process of finding the terms that multiply to give you a certain expression. One method for factoring is to find the **greatest common factor,** or **GCF.**

To find the GCF, list all the prime factors of each term and identify all common factors.

Example 11:

Find the GCF of each pair of monomials.

A. $4a^2$ and $10a$

B. $3q^5$, $12q^3$, and $18q^2$

Solution:

A. $4a^2 = \boxed{2} \cdot 2 \cdot \boxed{a} \cdot a$ List prime factors of each term

$10a = \boxed{2} \cdot 5 \cdot \boxed{a}$ Identify common factors

GCF: $2a$

B. $3q^5 = \boxed{3} \cdot q \cdot q \cdot q \cdot \boxed{q} \cdot \boxed{q}$

$12q^3 = 2 \cdot 2 \cdot \boxed{3} \cdot q \cdot \boxed{q} \cdot \boxed{q}$ List prime factors of each term

$18q^2 = 2 \cdot 3 \cdot \boxed{3} \cdot \boxed{q} \cdot \boxed{q}$ Identify common factors.

GCF: $3q^2$

Factor using GCF

To factor a polynomial, you can factor out the GCF to rewrite the expression as the product of its factors.

Example 12:

Factor $2x^3 - 6x^2 + 8x$.

Solution:

Step 1. Find the GCF.

$2x^3 = \boxed{2} \cdot \boxed{x} \cdot x \cdot x$	
$6x^2 = \boxed{2} \cdot 3 \cdot \boxed{x} \cdot x$	List prime factors of each term
$8x = \boxed{2} \cdot 2 \cdot 2 \cdot \boxed{x}$	Identify common factors

GCF: $2x$

Step 2. Factor out the GCF from each term.

$2x(x^2)+2x(-3x)+2x(4)$	Write as product of GCF and remaining terms
$2x(x^2-3x+4)$	Use distributive property to factor out the GCF

Factor by grouping

Recall that when using the distributive property to multiply binomials, you take each term of the first binomial and distribute each term into the second binomial separately.

$$(\underline{x}^2 + \underline{4})(x-7)$$

$$\swarrow \qquad \searrow$$

$$\underline{x}^2(x-7) + \underline{4}(x-7)$$

$$x^3 - 7x^2 + 4x - 28$$

Notice that after you multiply, you are left with four different terms. Using the distributive property to undo this process is called **factoring by grouping.**

Since x^2 was distributed into both terms in the binomial $(x - 7)$, you know that x^2 is a factor of both x^3 and $7x^2$.

Since 4 was distributed into both terms in the binomial $(x-7)$, you know that 4 is a factor of both $4x$ and 28.

$$x^3 - 7x^2 \quad + \quad 4x - 28$$

$$\text{GCF: } x^2 \qquad \text{GCF: } 4$$

Example 13:

Factor by grouping: $x^3 + 11x^2 + 6x + 66$.

Solution:

Step 1. With the polynomial in standard form, split the four terms into pairs and find the GCF of each pair of terms.

$$x^3 + 11x^2 + 6x + 66$$

$$x^3 = \boxed{x} \cdot \boxed{x} \cdot x \qquad 6x = \boxed{2} \cdot \boxed{3} \cdot x$$

$$11x^2 = 11 \cdot \boxed{x} \cdot \boxed{x} \qquad 66 = \boxed{2} \cdot \boxed{3} \cdot 11$$

$$\text{GCF: } x^2 \qquad \text{GCF: } 6$$

Step 2. Use the distributive property to factor out the GCF from each pair of terms.

$$x^3 + 11x^2 + 6x + 66$$

$$x^2(x+11) \qquad 6(x+11)$$

Notice that x^2 was distributed into $(x + 11)$ and then 6 was distributed into the same binomial, $(x + 11)$.

Step 3. Combine the GCFs into one binomial and rewrite as the product of both binomials.

$$(x^2 + 6)(x + 11)$$

Factoring Trinomials

Factoring by grouping is useful when trying to find the factors of a polynomial with four terms. When multiplying binomials using the distributive property, sometimes two of those terms can be combined into one term and simplified into a trinomial in the form $ax^2 + bx + c$.

You can also factor a trinomial in the form $ax^2 + bx + c$ by writing it as the product of two binomials.

$$(x+2)(x+3)$$

$$x^2 + 2x + 3x + 6$$

$$\searrow\swarrow$$

$$x^2 + 5x + 6$$

The product of the two binomials $(x + 2)$ and $(x + 3)$ is simplified to a trinomial in the form $ax^2 + bx + c$, where $a = 1$, $b = 5$, and $c = 6$.

The b term, 5, is the sum of 2 and 3.

The c term, 6, is the product of 2 and 3.

Factoring $x^2 + bx + c$

To factor a trinomial in the form $x^2 + bx + c$, look for numbers that have a sum of b and a product of c.

Example 14:

Factor $g^2 + 7g + 12$.

Solution:

Step 1. Find all factor pairs of c, 12.

1 and 12   2 and 6   3 and 4

Step 2. Choose the factor pair that has a sum of b, 7.

$1+12=13 \qquad 2+6=8 \qquad 3+4=\boxed{7}$

Step 3. Rewrite the trinomial as a four-term polynomial using the factor pair with a sum of b, 7, as coefficients of the linear terms.

$$g^2 + 3g + 4g + 12$$

Step 4. Factor by grouping.

$$g^2+3g+4g+12$$

$$g^2=\boxed{g}\cdot g \qquad 4g=\boxed{2}\cdot\boxed{2}\cdot g$$

$$3g=3\cdot\boxed{g} \qquad 12=\boxed{2}\cdot\boxed{2}\cdot 3$$

GCF: g GCF: 4

Step 5. Rewrite as the product of two binomials using the distributive property.

$$\underline{g}(g+3)+\underline{4}(g+3)$$

$$\searrow \quad \swarrow$$

$$(g+4)(g+3)$$

Factoring $x^2 - bx + c$

Recall that the product of two negative numbers is positive. Since you are looking for factors of *c* that add up to *b*, if *b* is negative, find negative factor pairs of *c*.

Example 15:

Factor $c^2 - 14c + 45$.

Solution:

Step 1. Find all *negative* factor pairs of 45.

$$-1 \text{ and } -45 \qquad -3 \text{ and } -15 \qquad -5 \text{ and } -9$$

Step 2. Choose the factor pair that has a sum of *b*, –14.

$$-1+(-45)=-46 \qquad -3+(-15)=-18 \qquad -5+(-9)=\boxed{-14}$$

Step 3. Rewrite the trinomial as a four-term polynomial using the factor pair with a sum of *b*, –14 as coefficients of the linear terms.

$$c^2 - 5c - 9c + 45$$

Step 4. Factor by grouping.

$$c^2 - 5c - 9c + 45$$

$$c^2 = \boxed{c} \cdot c \qquad -9c = \boxed{3} \cdot \boxed{3} \cdot -c$$

$$-5c = -5 \cdot \boxed{c} \qquad 45 = \boxed{3} \cdot \boxed{3} \cdot 5$$

GCF: c GCF: 9

Step 5. Rewrite as the product of two binomials using the distributive property.

$$c(c-5) - 9(c-5)$$

$$\searrow \quad \swarrow$$

$$(c-9)(c-5)$$

Factoring $x^2 + bx - c$

Recall that a negative product has one negative and one positive factor. Since you are looking for factors of *c* that add up to *b*, if *c* is negative, find one negative and one positive factor for the factor pairs of *c*.

Example 16:

Factor $m^2 + 14m - 32$.

Solution:

Step 1. Find all factor pairs of –32.

$$-1 \text{ and } 32 \qquad -2 \text{ and } 16 \qquad -4 \text{ and } 8$$

$$1 \text{ and } -32 \qquad 2 \text{ and } -16 \qquad 4 \text{ and } -8$$

Step 2. Choose the factor pair that has a sum of b, 14.

$$-1+32=31 \qquad -2+16=\boxed{14} \qquad -4+8=4$$

$$1+(-32)=-31 \qquad 2+(-16)=-14 \qquad 4+(-8)=-4$$

Step 3. Rewrite the trinomial as a four-term polynomial using the factor pair with a sum of b, 14.

$$m^2 - 2m + 16m - 32$$

Step 4. Factor by grouping.

$$m^2 - 2m + 16m - 32$$

$$m^2 = \boxed{m} \cdot m \qquad 16m = \boxed{2}\cdot\boxed{2}\cdot\boxed{2}\cdot\boxed{2}\cdot m$$

$$-2m = -2 \cdot \boxed{m} \qquad -32 = \boxed{2}\cdot\boxed{2}\cdot\boxed{2}\cdot\boxed{2}\cdot -2$$

GCF: m GCF: 16

Step 5. Rewrite as the product of two binomials using the distributive property.

$$m(m-2)+16(m-2)$$

$$\searrow \qquad \swarrow$$

$$(m+16)(m-2)$$

Factoring $ax^2 + bx + c$

Trinomials with a leading coefficient greater than 1 are the products of two binomials that also have one or more leading coefficients greater than 1.

$$(\underline{2}x + 7)(\underline{3}x + 4)$$

$(2x+7)(3x+4)$	Multiply using FOIL
$2x(3x)+2x(4)+7(3x)+7(4)$	
$6x^2+8x+21x+28$	Simplify products
$\boxed{6}x^2+29x+28$	Combine like terms

Notice that the leading coefficients of the two binomials create a leading coefficient greater than 1, and the leading coefficient of the trinomial, a, is the product of those two values.

$$(2x+7)(3x+4)$$
$$\boxed{6}x^2+29x+28$$

When you factor trinomials in the form x^2+bx+c, you look for factors of c that have a sum of b.

When factoring trinomials in the form ax^2+bx+c, the leading coefficient a changes the value of the two like terms that have a sum of b.

To factor a trinomial in the form ax^2+bx+c, find the factors of $a(c)$ that have a sum of b.

Example 17:

Factor $2p^2+5p+3$.

Solution:

Step 1. Find factor pairs of $a(c)$.

$$a=2 \text{ and } c=3$$

$$a(c)=2(3)=6$$

Find the factor pairs of 6.

1 and 6 2 and 3

Step 2. Choose the factor pair that has a sum of b, 5.

$1+6=7$ $2+3=\boxed{5}$

Step 3. Rewrite the trinomial as a four-term polynomial using the factor pair with a sum of b, 5.

$$2p^2 + 2p + 3p + 3$$

Step 4. Factor by grouping.

$$2p^2 + 2p + 3p + 3$$

$$2p^2 = \boxed{2} \cdot \boxed{p} \cdot p \qquad 3p = \boxed{3} \cdot p$$

$$2p = \boxed{2} \cdot \boxed{p} \qquad 3 = \boxed{3}$$

$$\text{GCF: } 2p \qquad \text{GCF: } 3$$

Step 5. Rewrite as the product of two binomials using the distributive property.

$$2p(p+1) + 3(p+1)$$

$$\searrow \quad \swarrow$$

$$(2p+3)(p+1)$$

Example 18:

Factor $8p^2 + 26p + 15$.

Solution:

Step 1. Find factor pairs of $a(c)$.

$$a = 8 \text{ and } c = 15$$

$$a(c) = 8(15) = 120$$

Find the factor pairs of 120.

1 and 120	2 and 60	3 and 40	4 and 30
5 and 24	6 and 20	8 and 15	10 and 12

Step 2. Choose the factor pair that has a sum of b, 26.

$1+120=121$	$2+60=62$	$3+40=43$	$4+30=34$
$5+24=29$	$6+20=\boxed{26}$	$8+15=23$	$10+12=22$

Step 3. Rewrite the trinomial as a four-term polynomial using the factor pair with a sum of b, 5.

$$8p^2 + 6p + 20p + 15$$

Step 4. Factor by grouping.

$$8p^2 + 6p + 20p + 15$$

$$8p^2 = 2 \cdot 2 \cdot \boxed{2} \cdot \boxed{p} \cdot p \qquad 20p = 2 \cdot 2 \cdot \boxed{5} \cdot p$$

$$6p = 3 \cdot \boxed{2} \cdot \boxed{p} \qquad 15 = 3 \cdot \boxed{5}$$

GCF: $2p$ GCF: 5

Step 5. Rewrite as the product of two binomials using the distributive property.

$$2p(4p+3) + 5(4p+3)$$

↘ ↙

$$(2p+5)(4p+3)$$

Perfect square trinomials

Recall that there are squares of binomials that create perfect square trinomials.

Square of a binomial: $(a + b)^2 = a^2 + 2ab + b^2$

Difference of squares: $(a + b)(a - b) = a^2 - b^2$

Square of a binomial

Example 19:

Factor $x^2 - 6x + 9$.

Solution:

Step 1. Find factor pairs of c. Since b is negative and c is positive, both factors are negative.

-1 and -9 -3 and -3

Step 2. Choose the factor pair that has a sum of b, -6.

$$-1 + (-9) = -10 \qquad -3 + (-3) = \boxed{-6}$$

Step 3. Rewrite the trinomial as a four-term polynomial using the factor pair with a sum of b, –6.

$$x^2 - 3x - 3x + 9$$

Step 4. Factor by grouping.

$$x^2 - 3x - 3x + 9$$

$x^2 = x \cdot \boxed{x}$	$-3x = \boxed{3} \cdot -x$
$-3x = -3 \cdot \boxed{x}$	$9 = \boxed{3} \cdot 3$
GCF: x	GCF: 3

Step 5. Rewrite as the product of two binomials using the distributive property.

$$x(x-3) - 3(x-3)$$

$$(x-3)(x-3)$$

Step 6. Rewrite as the square of a binomial.

$$(x - 3)^2$$

Difference of squares

Example 20:

Factor $x^2 - 49$.

Solution:

Step 1. Find factor pairs of c. Since c is negative, one factor is negative and the other is positive.

-1 and 49 $\qquad$ 1 and -49 $\qquad$ -7 and 7

Step 2. Choose the factor pair that has a sum of b. Since there is no b term, the sum is 0.

$-1 + 49 = 48$ $\qquad$ $1 + (-49) = -48$ $\qquad$ $-7 + 7 = \boxed{0}$

Step 3. Rewrite the trinomial as a four-term polynomial using the factor pair with a sum of *b*, 0.

$$x^2 - 7x + 7x + 49$$

Step 4. Factor by grouping.

$$x^2 - 7x + 7x + 49$$

$x^2 = x \cdot \boxed{x}$	$7x = \boxed{7} \cdot x$
$-7x = -7 \cdot \boxed{x}$	$49 = \boxed{7} \cdot 7$
GCF: x	GCF: 7

Step 5. Rewrite as the product of two binomials using the distributive property.

$$x(x+7) - 7(x+7)$$

$$\searrow \quad \swarrow$$

$$(x-7)(x+7)$$

Chapter Check-Out

Questions

For questions 1–2, find the degree of each monomial.

1. 14
2. $-10x^2y^9$

For questions 3–4, write each polynomial in standard form and name it based on its degree and its number of terms.

3. $4t^2 + 19j - 2t^2$
4. $6b^3 + 7c^2 - b^3 + 8$
5. Find the perimeter of the following figure.

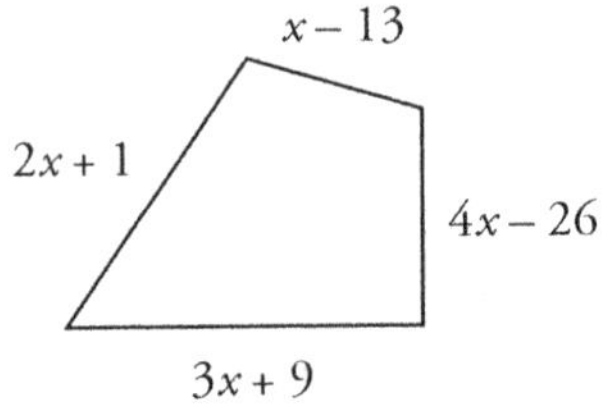

For questions 6–7, simplify the sum or difference of the polynomials.

6. $(3x^2 + x - 5) + (4x^2 - 2x + 8)$
7. $(6n^2 - 5n + 2) - (n^2 - 8n + 3)$

For questions 8–10, simplify the product.

8. $-3z(z^3 - 4z + 6)$
9. $(2d - 5)(d + 11)$
10. $(x - 2)(7x^2 + 3x - 1)$
11. Find the area of the following rectangle.

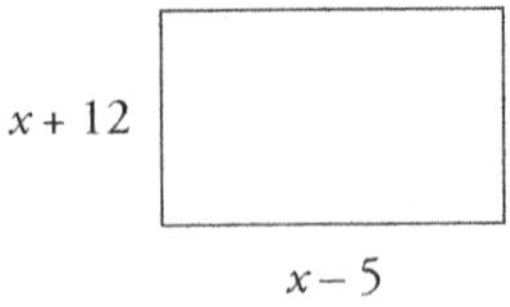

For questions 12–13, factor out the GCF.

12. $6w^3 + 9w$
13. $-12p^2 + 8$

For questions 14–15, factor by grouping.

14. $x^3 + 7x^2 - 5x - 35$

15. $h^3 - 2h^2 + 8h - 16$

For questions 16–21, factor the expression.

16. $x^2 + 9x + 20$

17. $k^2 - 18k + 77$

18. $y^2 - 5y - 24$

19. $3p^2 - 2p - 5$

20. $t^2 + 6t + 9$

21. $h^2 - 169$

Answers

1. 0
2. 11
3. $2t^2 + 19j$; quadratic binomial
4. $5b^3 + 7c^2 + 8$; cubic trinomial
5. $10x - 29$
6. $7x^2 - x + 3$
7. $5n^2 + 3n - 1$
8. $-3z^4 + 12z^2 - 18z$
9. $2d^2 + 17d - 55$
10. $7x^3 - 11x^2 - 7x + 2$
11. $x^2 + 7x - 60$
12. $3w(2w^2 + 3)$
13. $4(-3p^2 + 2)$
14. $(x^2 - 5)(x + 7)$
15. $(h^2 + 8)(h - 2)$
16. $(x + 4)(x + 5)$
17. $(k - 7)(k - 11)$
18. $(y + 3)(y - 8)$
19. $(3p - 5)(p + 1)$
20. $(t + 3)^2$
21. $(h + 13)(h - 13)$

Chapter 9

QUADRATIC FUNCTIONS

Chapter Check-In

- ❑ Parabolas
- ❑ Effects on graph of a quadratic parent function
- ❑ Vertex form of a quadratic function
- ❑ Solving quadratic equations by graphing
- ❑ Solving quadratic equations by finding the square roots
- ❑ Solving quadratic equations by factoring
- ❑ Solving quadratic equations that cannot be factored: Completing the square and the quadratic formula
- ❑ TECH TIPS: Solving quadratic equations on a graphing calculator

Any function that can be written in the **standard form,** $y = ax^2 + bx + c$, where $a \neq 0$, is a **quadratic function.** The **quadratic parent function,** $f(x) = x^2$, is the simplest form of a quadratic function.

Parabolas

Let's evaluate the quadratic parent function, $f(x) = x^2$, for a domain of $\{-2, -1, 0, 1, 2\}$.

x	x^2	y
–2	$(-2)^2$	4
–1	$(-1)^2$	1
0	0^2	0
1	1^2	1
2	2^2	4

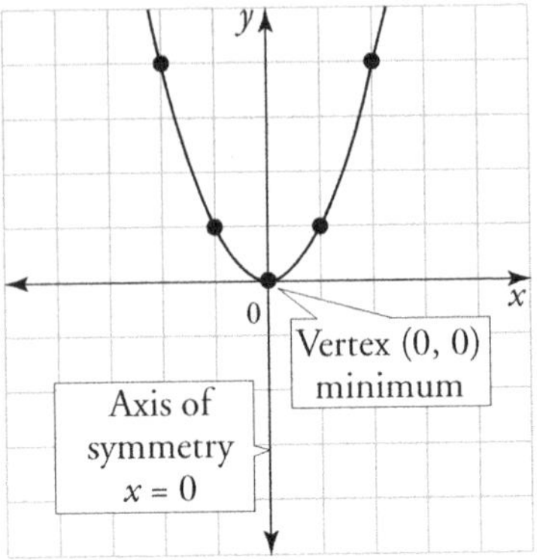

Notice that the graph of a quadratic function creates a smooth u-shaped curve called a **parabola.**

Observe that the output values decrease to a **minimum** value, called the **vertex,** and then increase again, repeating the same output values. This causes the graph of the quadratic parent function to have **symmetry** about the y-axis, which has an equation of $x = 0$. In the quadratic parent function, we say that the vertical line $x = 0$ is the **axis of symmetry.** The vertex lies on the axis of symmetry.

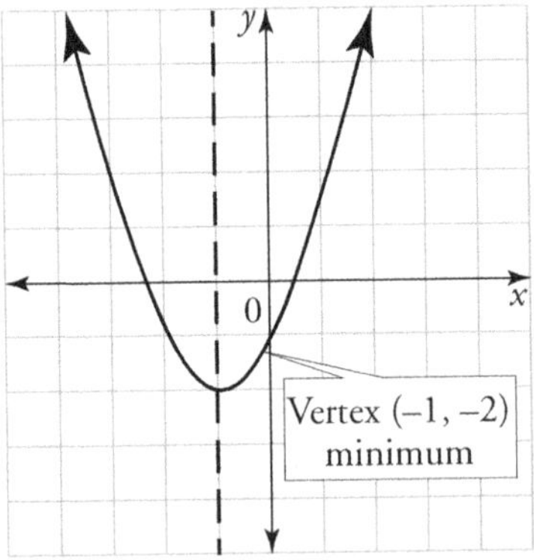

A parabola that *opens up* has a vertex that is a **minimum**.

axis of symmetry, $x = -1$

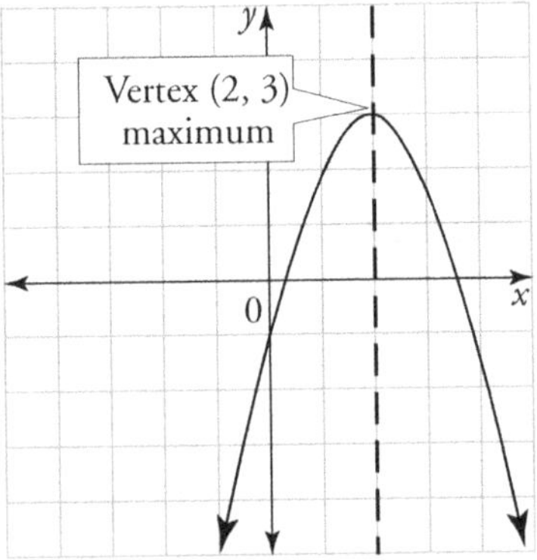

A parabola that *opens down* has vertex that is a **maximum**.

axis of symmetry, $x = 2$

Notice that both types of parabolas continue infinitely in both directions. The domain of a quadratic function is all real numbers, or $(-\infty, \infty)$.

The vertex serves as a boundary for the range values.

Range: $y \geq -2$
Range: $y \leq 3$

Example 1:

Identify the vertex and state if it is a minimum or a maximum. Write the equation for the axis of symmetry, and state the domain and range.

A.

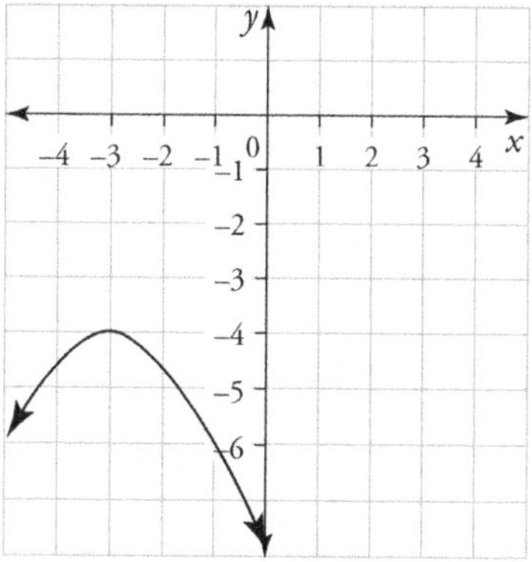

B.

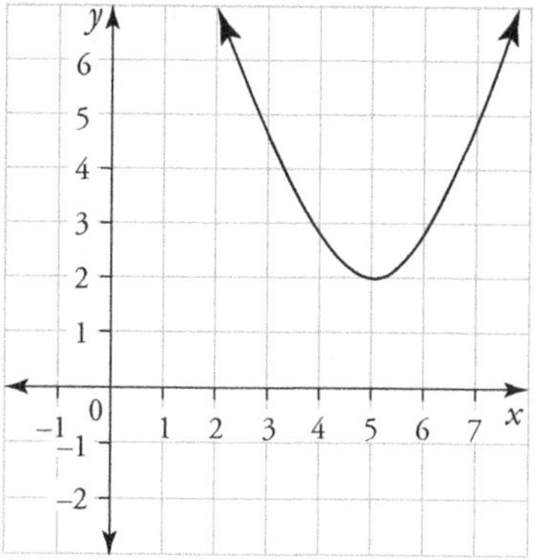

Solution:

A.
Vertex: $(-3, -4)$ *maximum*
axis of symmetry, $x = -3$
Domain: all real numbers
Range: $y \leq -4$

B.
Vertex: $(5, 2)$ *minimum*
axis of symmetry, $x = 5$
Domain: all real numbers
Range: $y \geq 2$

Effects on Graph of Quadratic Parent Function

Effects of $af(x)$

Observe what happens when you multiply the quadratic parent function $f(x) = x^2$ by some value, a.

$$a \cdot f(x) = a(x^2)$$

when $a = \frac{1}{2}$		
x	x^2	$\frac{1}{2}(x)^2$
–2	4	2
–1	1	$\frac{1}{2}$
0	0	0
1	1	$\frac{1}{2}$
2	4	2

when $a = 2$		
x	x^2	$2(x^2)$
–2	4	8
–1	1	2
0	0	0
1	1	2
2	4	8

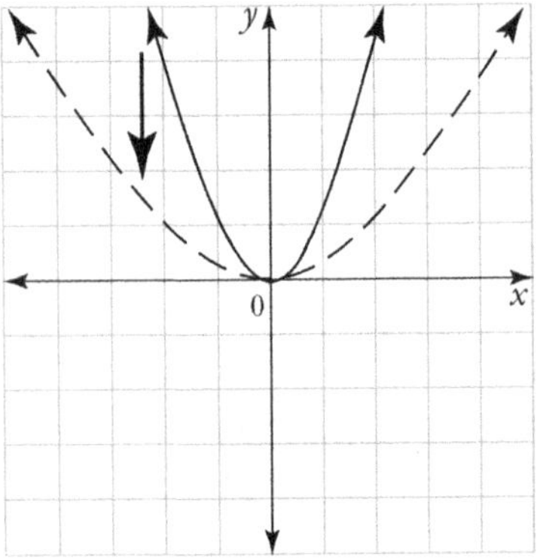

$0 < |a| < 1$

The parabola gets *wider*

vertical compression toward the x-axis

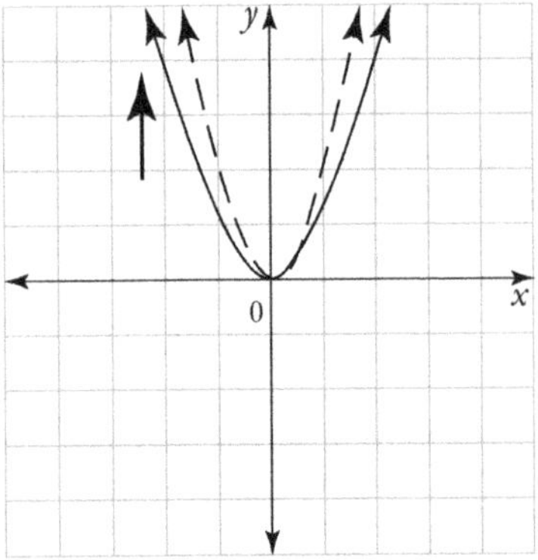

$|a| > 1$

The parabola gets *narrower*

vertical stretch away from the x-axis

when $a = -1$		
x	x^2	$-(x^2)$
−2	4	−4
−1	1	−2
0	0	0
1	1	−2
2	4	−4

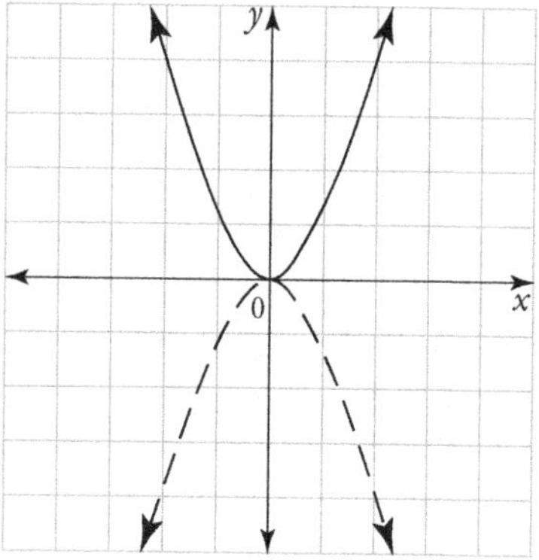

$a < 0$

reflection across the x-axis

Effects of $f(x) + k$

Observe what happens when you add some value, k, to the output value of the quadratic parent function $f(x) = x^2$.

$$f(x) + k = x^2 + k$$

when $k = 3$		
x	x^2	$x^2 + 3$
−2	4	7
−1	1	4
0	0	3
1	1	4
2	4	7

when $k = -1$		
x	x^2	$x^2 - 1$
−2	4	3
−1	1	0
0	0	−1
1	1	0
2	4	1

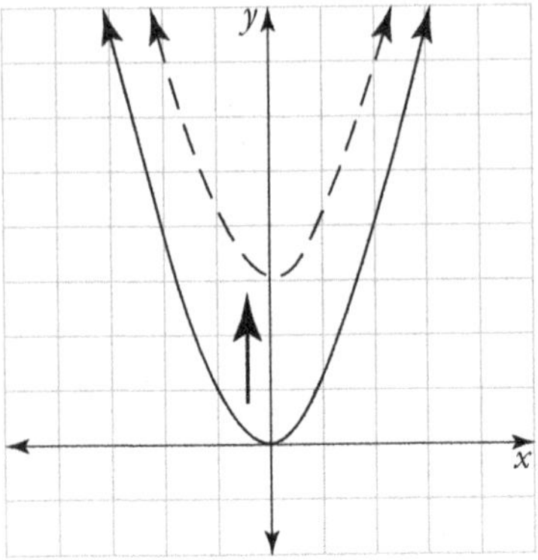

$k < 0$

Vertical translation of k units *up*

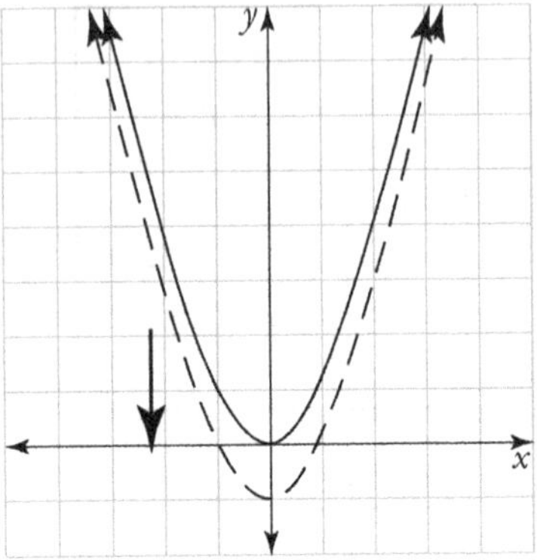

$k > 0$

Vertical translation of k units *down*

Notice that the coordinates of the new vertex can be described as

$$f(x) + k = y + k = x^2 + k$$

$$(x, y) \rightarrow (x, y + k)$$

Effects of $f(x + h)$

Observe what happens when you add some value, h, to the input value of the quadratic parent function $f(x) = x^2$.

$$f(x + h) = (x + h)^2$$

when $h = -1$		
x	$x - 1$	$(x - 1)^2$
–2	–3	9
–1	–2	4
0	–1	1
1	0	0
2	1	1

when $h = 4$		
x	$x + 4$	$(x + 4)^2$
–2	2	4
–1	3	9
0	4	16
1	5	25
2	6	36

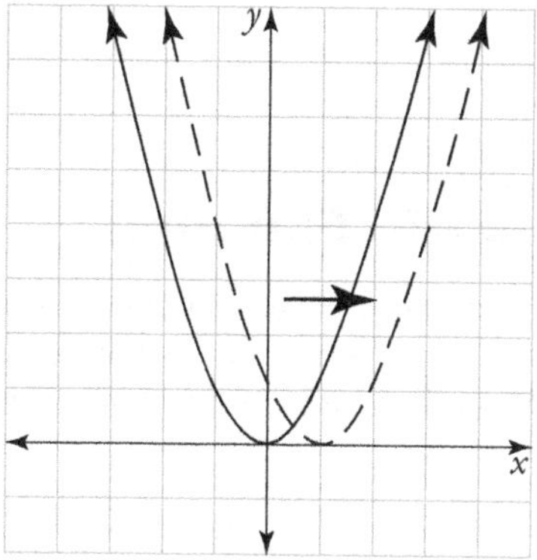

$h < 0$

Horizontal translation of h units *to the right*

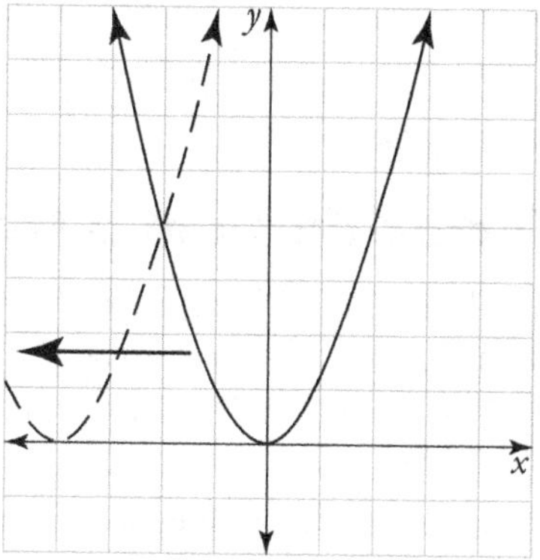

$h > 0$

Horizontal translation of h units *to the left*

Notice that the value of h moves the vertex in the opposite direction. Coordinates of the new vertex can be described as follows:

$$f(x + h) = (x + h)^2$$

$$(x, y) \rightarrow (x - h, y)$$

Example 2:

For each of the following, describe the transformation of the quadratic parent function. Identify the coordinates of the new vertex.

A. $f(x) = 3x^2 - 1$

B. $f(x) = -(x + 2)^2$

C. $f(x) = (x - 6)^2 + 5$

Solution:

A. $a = 3$, gets narrower, *vertical stretch*
$k = -1$, translated *down* 1 unit
new vertex: $(0, -1)$

B. $a = -1$, reflected over the x-axis
$h = 2$, translated *left* 2 units
new vertex: $(-2, 0)$

C. $h = -6$, translated right 6 units

$k = 5$, translated *up* 5 units

new vertex: (6, 5)

Vertex Form of a Quadratic Function

Recall that multiplying by some value, a, causes a vertical compression when $0 < a < 1$ or a vertical stretch when $a > 1$. The sign of a also describes if the parabola opens up (when a is positive) or down (when a is negative).

The quadratic parent function has a vertex at the origin, (0, 0). You saw in the transformations earlier in the chapter that inserting any value, h, as the input and any value, k, as the output creates a new vertex at $(-h, k)$. Remember, the value of h always moves the vertex in the opposite direction.

You can write a transformation of a quadratic parent function in **vertex form.**

Vertex Form of a Quadratic Function:

$$f(x) = a(x - h)^2 + k$$

where a, h, and k are all constants and the vertex is (h, k)

When given the vertex of a parabola, (h, k), you can write the equation of the parabola if you know one additional point, (x, y).

Example 3:

Write the equation of a parabola that has a vertex at (2, 1) and passes through the point (3, 4) in vertex form.

Solution:

Step 1. Substitute the coordinates of the vertex for (h, k).

$$y = a(x - h)^2 + k$$

$$y = a(x - 2)^2 + 1 \qquad \text{Substitute 2 for } h \text{ and substitute 1 for } k$$

Step 2. Substitute the coordinates of the point for (x, y).

$$4 = a(3 - 2)^2 + 1 \qquad \text{Substitute 3 for } x \text{ and 4 for } y$$

Step 3. Solve for *a*.

$4 = a(3-2)^2 + 1$	
$4 = a(1)^2 + 1$	Simplify subtraction inside parentheses
$4 = 1a + 1$	Simplify exponent
$3 = a$	Subtract 1 from both sides

Step 4. Write the equation in vertex form using vertex (*h*, *k*) and the value of *a*.

$$f(x) = 3(x-2)^2 + 1$$

You can use the order of operations and the distributive property to rewrite a quadratic equation in vertex form in standard form: $y = ax^2 + bx + c$.

Example 4:

Rewrite the equation $f(x) = 5(x+1)^2 - 10$ in standard form.

Solution:

Step 1. Use the order of operations to simplify exponents first.

$f(x) = 5(x+1)^2 - 10$	Rewrite as a square of binomials
$f(x) = 5(x+1)(x+1) - 10$	

Step 2. Simplify product of binomials using FOIL or the distributive property.

$f(x) = 5(x+1)(x+1) - 10$	Multiply using FOIL
$f(x) = 5(x^2 + 2x + 1) - 10$	Combine like terms

Step 3. Distribute the value of *a* into the parentheses and simplify.

$f(x) = 5(x^2 + 2x + 1) - 10$	
$f(x) = 5x^2 + 10x + 5 - 10$	Use distributive property
$f(x) = 5x^2 + 10x - 5$	Combine like terms

Solving Quadratic Equations by Graphing

The solutions to a quadratic function in the form $0 = ax^2 + bx + c$ are the x-intercepts of its graph.

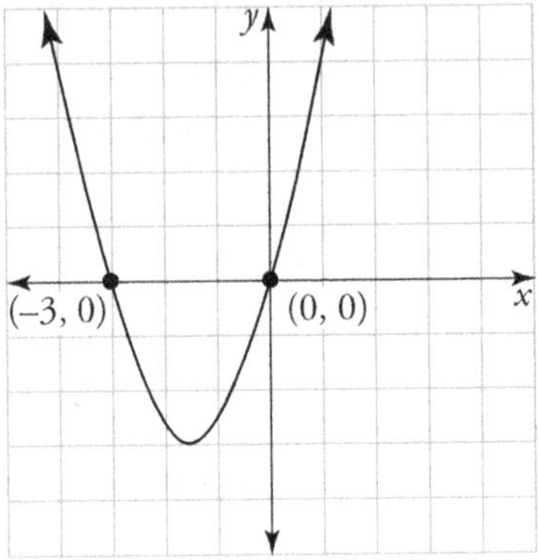

Since all x-intercepts have a y-value of 0, they can also be referred to as the **zeros of the function.** The solution of a quadratic function can be thought of as the value of x that makes $y = 0$.

> **Quadratic Equation:**
>
> An equation written in the form $ax^2 + bx + c = 0$, where $a \neq 0$.

Sometimes you can identify the solutions to a quadratic equation in the form $ax^2 + bx + c = 0$ from a table or graph.

Example 5:

Solve by making a table of values and graphing.

A. $x^2 = 0$

B. $x^2 - 1 = 0$

C. $x^2 + 2 = 0$

Solution:

A. $x^2 = 0$

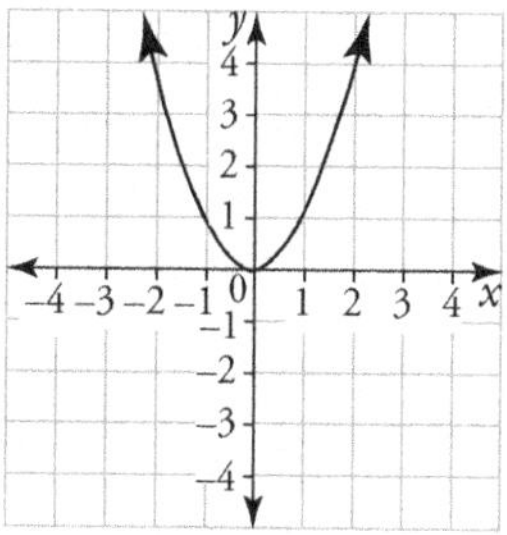

x	y
–2	4
–1	1
0	0
1	1
2	4

One solution, 0

B. $x^2 - 1 = 0$

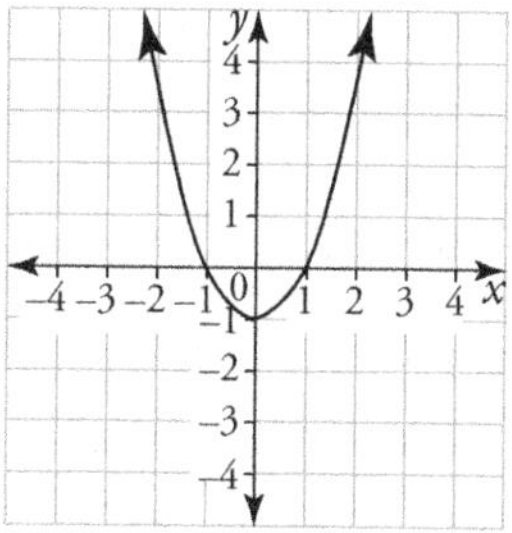

x	y
–2	3
–1	0
0	–1
1	0
2	3

Two solutions, ±1

C. $x^2 + 2 = 0$

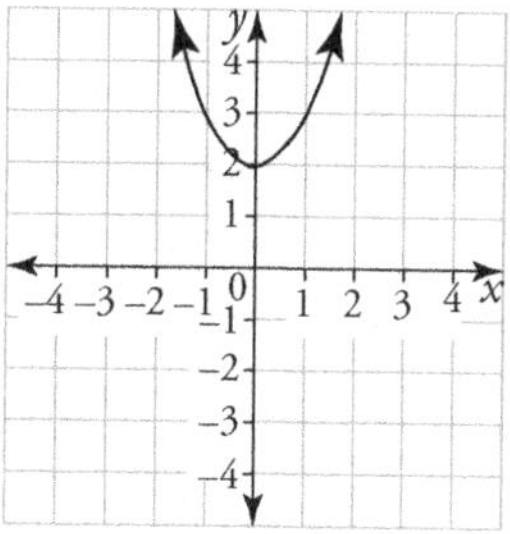

x	y
–2	6
–1	3
0	2
1	3
2	6

No solution

Solving Quadratic Equations by Finding the Square Roots

The solutions of a quadratic equation are also known as **roots** since equations in the form $k = x^2$ for some constant k, can be solved by performing the inverse operation, finding a square root.

Example 6:

Solve by finding the square roots.

A. $4x^2 = 0$

B. $2x^2 - 72 = 0$

C. $x^2 + 49 = y$

Solution:

A. $4x^2 = 0$

$\frac{4x^2}{4} = \frac{0}{4}$ Divide both sides by 4

$x^2 = 0$ Simplify

$x = \sqrt{0}$ Find the square roots

$x = 0$ One solution, $x = 0$

B. $2x^2 - 72 = 0$

$2x^2 = 72$ Add 72 to both sides

$x^2 = 36$ Divide both sides by 2

$x = \pm\sqrt{36}$ Find the square roots

$x = \pm 6$ Two solutions, ± 6

C. $x^2 + 49 = y$

$x^2 + 49 = 0$ Substitute 0 for y

$x^2 = -49$ Subtract 49 from both sides

$x = \sqrt{-49}$ Find the square roots

Note that there is no number in the real number system that can be squared to give a negative product. Therefore, there is *no real solution.*

Solving Quadratic Equations by Factoring

Recall from Chapter 8 that some quadratic polynomials in the form $ax^2 + bx + c$ can be rewritten as the product of its linear factors.

You can use the **zero-product property** to find the solutions of a quadratic equation set equal to 0 and rewritten as the product of its linear factors.

Zero-Product Property:

If $ab = 0$, then $a = 0$ or $b = 0$.

Example 7:

Solve $(3x - 6)(x + 4) = 0$.

Solution:

	$3x-6=0$	or	$x+4=0$	Use zero-product property
Add 6 to both sides	$3x=6$		$x=-4$	Subtract 4 from both sides
Divide by 3	$x=2$			
	$x=2$	or	$x=-4$	

Example 8:

Solve $x^2 + 4x - 21 = 0$.

Solution:

Step 1. Factor $x^2 + 4x - 21$.

Find factors of –21, and pick the factors with a sum of 4.

–1 and 21	–3 and 7 (circled)
3 and 7	1 and 21

$x^2 - 3x + 7x - 21$	Rewrite *b* as the sum of factors
$x(x-3) + 7(x-3)$	Factor out the GCF
$(x+7)(x-3)$	Rewrite as product of binomials

Step 2. Use zero-product property.

$$(x+7)(x-3) = 0$$

$x + 7 = 0$ or $x - 3 = 0$ Solve for x

Subtract 7	$x+7=0$		$x-3=0$	Add 3
	$x=-7$	or	$x=3$	

Example 9:

Solve $5x^2 + 21x = -4$.

Step 1. Set equal to zero.

$$5x^2 + 21x + 4 = 0 \qquad \text{Add 4 to both sides}$$

Step 2. Factor $5x^2 + 21x + 4$.

$$(5x+1)(x+4) \qquad \text{Rewrite as product of binomials}$$

Step 3. Use zero-product property.

$$(5x+1)(x+4) = 0$$

$$5x + 1 = 0 \quad \text{or} \quad x + 4 = 0$$

Subtract 1 $\quad 5x = -1 \qquad x = -4 \quad$ Subtract 4

Divide by 5 $\quad x = -\frac{1}{5}$

$$x = -\frac{1}{5} \quad \text{or} \quad x = -4$$

Solving Quadratic Equations that Cannot Be Factored

Some quadratic functions cannot be rewritten as the product of two binomials. Let's explore additional methods to solve quadratic equations that cannot be factored.

Completing the square

Recall that perfect square trinomials are the product of binomials squared.

$$(x + \boxed{4})^2 = (x + 4)(x + 4) = x^2 + 4x + 4x + 16$$

$$x^2 + \boxed{8x} + \boxed{16}$$

$$b = 2(4) \qquad c = 4^2$$

A perfect square trinomial with $a = 1$ is written in the form $ax^2 + bx + c$, where

$$c = \left(\frac{b}{2}\right)^2$$

Perfect Square Trinomial:

$$x^2 + bx + \left(\frac{b}{2}\right)^2$$

Example 10:

Find the value of c that makes $x^2 + 6x + c$ a perfect square trinomial.

Solution:

$$x^2 + 6x + \left(\frac{6}{2}\right)^2$$

$$x^2 + 6x + 3^2 = x^2 + 6x + \boxed{9}$$

$$c = 9$$

Example 11:

Solve $p^2 - 2p = 8$ by completing the square.

Solution:

Step 1. Find the value of c that makes $p^2 - 2p + c$ a perfect square.

$$\left(\frac{b}{2}\right)^2 = \left(\frac{-2}{2}\right)^2 = (-1)^2$$

$$c = 1$$

Step 2. Rewrite the left side of the equation as a perfect square trinomial by adding c to both sides.

$p^2 - 2p = 8$	$c = 1$
$p^2 - 2p \boxed{+1} = 8 \boxed{+1}$	Add 1 to both sides
$p^2 - 2p + 1 = 9$	Simplify

Step 3. Factor the perfect square trinomial on the left side of the equation.

$$(p - 1)^2 = 9$$

Step 4. Solve.

$p - 1 = \pm\sqrt{9}$	Find the square roots
$p - 1 = \pm 3$	Simplify
$p - 1 = 3 \quad \text{or} \quad p - 1 = -3$	Rewrite as two equations
$p = 4 \quad \text{or} \quad p = -2$	Add 1 to both sides

Example 12:

Solve $w^2 + 12w + 20 = 0$ by completing the square.

Solution:

Step 1. Rewrite as $x^2 + bx = k$

$w^2 + 12w = -20$	Subtract 20 from both sides

Step 2. Find the value of c that makes $w^2 + 12w + c$ a perfect square.

$$\left(\frac{b}{2}\right)^2 = \left(\frac{12}{2}\right)^2 = 6^2$$

$$c = 36$$

Step 3. Rewrite the left side of the equation as a perfect square trinomial by adding *c* to both sides.

$w^2 + 12w = -20$	$c = 36$
$w^2 + 12w \boxed{+36} = -20 \boxed{+36}$	Add 36 to both sides
$w^2 + 12w + 36 = 16$	Simplify

Step 4. Factor the perfect square trinomial on the left side of the equation.

$$(w + 6)^2 = 16$$

Step 5. Solve.

$w + 6 = \pm\sqrt{16}$	Find the square roots
$w + 6 = \pm 4$	Simplify
$w + 6 = 4 \quad \text{or} \quad w + 6 = -4$	Rewrite as two equations
$w = -2 \quad \text{or} \quad w = -10$	Subtract 6 from both sides

Quadratic formula

You can also solve any quadratic equation in the form $ax^2 + bx + c$ by using the **quadratic formula.**

Quadratic Formula:

$$x = \frac{-b \pm \sqrt{b^2 - 4ac}}{2a}$$

Example 13:

Solve $n^2 + 9n + 14 = 0$ using the quadratic formula.

Solution:

Step 1. Substitute values of a, b, and c into the quadratic formula.

$$n = \frac{-(9) \pm \sqrt{(9)^2 - 4(1)(14)}}{2(1)}$$ Substitute 1 for a, 9 for b, and 14 for c

$$n = \frac{-9 \pm \sqrt{81 - 56}}{2}$$ Use order of operations

$$n = \frac{-9 \pm \sqrt{25}}{2} = \frac{-9 \pm 5}{2}$$ Simplify the radicand

Step 2. Rewrite as two equations.

$$n = \frac{-9 + 5}{2} \quad \text{or} \quad n = \frac{-9 - 5}{2}$$

$$n = -2 \quad \text{or} \quad n = -7$$ Simplify

Some solutions can be approximated using the quadratic formula and the square root function on your calculator.

Example 14:

Solve $3t^2 - 7t - 15 = 0$ using the quadratic formula.

Solution:

Step 1. Substitute values of a, b, and c into the quadratic formula.

$$t = \frac{-(-7) \pm \sqrt{(-7)^2 - 4(3)(-15)}}{2(3)}$$ Substitute 3 for a, -7 for b, and -15 for c

$$t = \frac{7 \pm \sqrt{49 + 180}}{6}$$ Use order of operations

$$t = \frac{7 \pm \sqrt{229}}{6}$$ Simplify the radicand

$$t \approx \frac{7 \pm 15.13}{6}$$ Approximate $\sqrt{229}$ on calculator

Step 2. Rewrite as two equations.

$$t \approx \frac{7 + 15.13}{6} \quad \text{or} \quad t \approx \frac{7 - 15.13}{6}$$

$$t \approx 3.69 \quad \text{or} \quad t \approx -1.36$$ Simplify

Example 15:

Solve $2a^2 + 3a = -5$.

Solution:

Step 1. Rewrite in standard form.

$$2a^2 + 3a + 5 = 0$$ Add 5 to both sides

Step 2. Substitute values of a, b, and c into the quadratic formula.

$$a=\frac{-(3)\pm\sqrt{(3)^2-4(2)(5)}}{2(2)}$$ Substitute 2 for a, 3 for b, and 5 for c

$$a=\frac{-3\pm\sqrt{9-40}}{4}$$ Use order of operations

$$a=\frac{-3\pm\sqrt{-31}}{4}$$ Simplify the radicand

Because the radicand is negative, there is *no real solution.*

TECH TIPS: Solving Quadratic Equations on a Graphing Calculator

Recall that the solution to a quadratic function is sometimes called a root, a zero, or an x-intercept. The solution is the value of x that makes $ax^2 + bx + c = 0$.

To find that value of x that makes $y = 0$, let's first look at the table of values.

Solving using a table

Example:

Solve $x^2 + 2x - 8 = 0$.

Solution:

1. Press [Y=] and enter the expression on the left into Y_1 =.

Plot 1	Plot 2	Plot 3
\Y_1= $X^2 + 2X - 8$		
\Y_2=		
\Y_3=		
\Y_4=		
\Y_5=		
\Y_6=		
\Y_7=		

2. Press **2nd** **GRAPH** to view the table.

   The *y*-values for the expression on the left are listed under Y_1.

3. Press the UP or DOWN arrow keys to look for the value of *x* that makes $Y_1 = 0$.

X	Y_1	Y_2
-4	0	
-3	-5	
-2	-8	
-1	-9	
0	-8	
1	-5	
2	0	
X=-4		

Answer: $x = -4$ or $x = 2$

Solving using a graph

When solutions to a quadratic function are not whole number values, it's helpful to use a graph to solve.

Example:

Solve $x^2 + 4x - 6 = 0$.

Solution:

Any equation can be viewed as two separate expressions on opposite sides of the equal sign.

$$x^2 + 4x - 6 = 0$$

Expression on left	and	Expression on right
$x^2 + 4x - 6$		0

1. Press **Y =** and enter the expression on the left into Y_1 =.

   Enter the expression on the right into Y_2 =.

Plot 1	Plot 2	Plot 3
\Y₁= X²+4X−6		
\Y₂= 0		
\Y₃=		
\Y₄=		
\Y₅=		
\Y₆=		
\Y₇=		

2. Press **2nd** **GRAPH** to view the table.

   The *y*-values for the expression on the *left* are listed under Y_1.

   Look for the value of *x* that will make $Y_1 = 0$.

3. Notice that Y_1 would equal 0 for a value of *x* that's between 1 and 2.

   When the solution is not visible on the table, it is helpful to use the graph.

X	Y_1	Y_2
-2	-10	0
-1	-9	0
0	-6	0
1	-1	0
2	6	0
3	15	0
4	26	0
X=1		

4. Press **GRAPH**

   Notice that the *x*-intercepts appear to be near –5 and 1.

   The *x*-axis has an equation of $y = 0$.

5. Press **2nd** **TRACE** to calculate where the parabola intersects with the *x*-axis at $y = 0$.

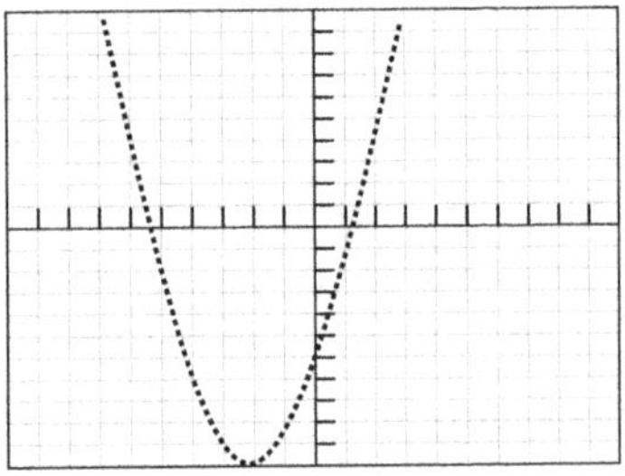

6. While in CALC mode, press **5** to calculate the point of intersection.

7. The screen will prompt you with three questions:

   First curve? Second curve? Guess?

   Move the cursor close to the first *zero* you wish to find.

   Press **ENTER** three times.

CALCULATE
1: value
2: zero
3: minimum
4: maximum
5: intersect
6: dy/dx
7: ∫f(x)dx

8. The point of intersection will appear at the bottom of the screen.

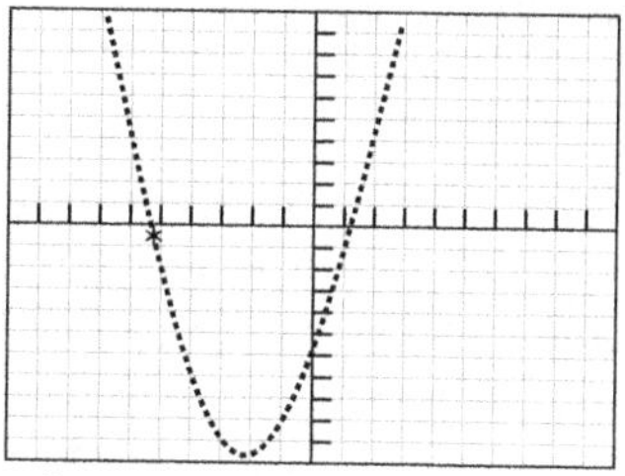

Intersection
X = –5.162278 Y = 0

9. Repeat Steps 5–7, placing the cursor near the second *zero*.

Answer: $x \approx -5.16$ and $x \approx 1.16$

Chapter Check-Out

Questions

For questions 1–2, identify the vertex and state if it is a minimum or a maximum. Write the equation for the axis of symmetry and state the domain and range.

1.

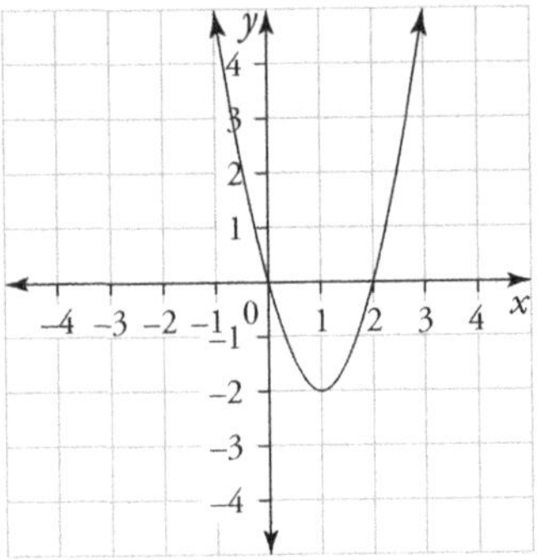

2.

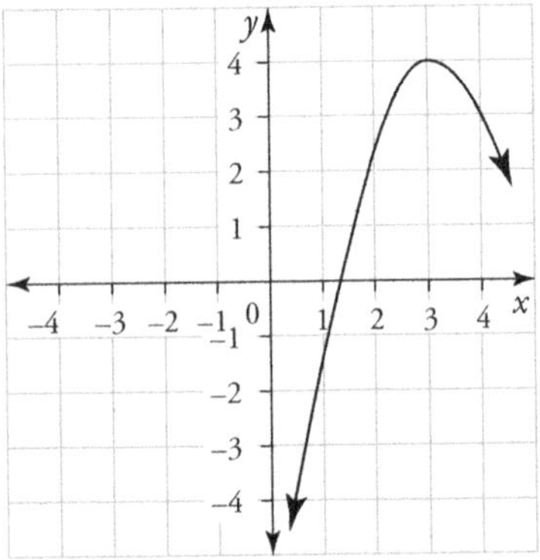

For questions 3–4, describe the transformation of the quadratic parent function. Identify the coordinates of the new vertex.

3. $f(x) = -2(x + 8)^2$

4. $f(x) = (x - 5)^2 + 3$

For questions 5–6, write the equation of a parabola with the given characteristics in vertex form.

5. vertex at (2, 3), passes through the point (4, 11)

6. vertex at (1, 0), passes through the point (2, 3)

7. Rewrite $f(x) = 2(x - 3)^2 + 4$ in standard form.

For questions 8–10, solve by making a table of values and graphing.

8. $0 = 2x^2 + 5$

9. $0 = (x - 3)^2$

10. $0 = x^2 - 4$

For questions 11–13, solve by finding the square roots.

11. $5x^2 = 0$

12. $3x^2 - 27 = 0$

13. $x^2 + 15 = 0$

For questions 14–15, solve using the zero-product property.

14. Solve $(2x + 4)(x - 1) = 0$.

15. Solve $6x^2 - 9x - 6 = 0$.

For questions 16–17, find the value of c that makes the trinomial a perfect square.

16. $x^2 - 8x + c$

17. $x^2 + 3x + c$

18. Solve $h^2 - 10h - 11 = 0$ by completing the square.

For questions 19–20, solve using the quadratic formula.

19. $r^2 + 3r - 5 = 0$

20. $2v^2 - 6v + 7 = 0$

Answers

1. vertex (1, –2), minimum;
   $x = 1$;
   Domain: all real numbers
   Range: $y \geq -2$
2. vertex (3, 4), maximum;
   $x = 3$;
   Domain: all real numbers
   Range: $y \leq 4$
3. reflected across x-axis,
   vertical stretch,
   translated 8 units left;
   (–8, 0)
4. translated 5 units right,
   3 units up;
   (5, 3)
5. $f(x) = 2(x - 2)^2 + 3$
6. $f(x) = 3(x - 1)^2$
7. $y = 2x^2 - 12x + 22$
8. no real solution
9. $x = 3$
10. $x = \pm 2$
11. $x = 0$
12. $x = \pm 3$
13. no real solution
14. $x = 1$ or $x = -2$
15. $x = 2$ or $x = -\frac{1}{2}$
16. $c = 16$
17. $c = \frac{9}{4}$
18. $x = 11$ or $x = -1$
19. $r \approx -4.19, 1.19$
20. no real solution

ALGEBRA I FORMULA CHART

General Formulas

Slope of a line	$m = \frac{y_2 - y_1}{x_2 - x_1}$
Pythagorean theorem	$a^2 + b^2 = c^2$
Quadratic formula	$x = \frac{-b \pm \sqrt{b^2 - 4ac}}{2a}$

Forms of Linear Equations

Slope-intercept form	$y = mx + b$
Point-slope form	$y - y_1 = m(x - x_1)$
Standard form	$Ax + By = C$

Geometric Formulas

Please note the following abbreviations used in geometric formulas:

Circumference:
C = circumference
r = radius
d = diameter
$\pi \approx 3.14$

Volume:
V = volume
B = area of base
h = height
r = radius
$\pi \approx 3.14$

Area:
A = area
b = base
h = height
d = diagonal
a = apothem
P = perimeter of base

Surface Area:
S = surface area
P = perimeter of base
l = slant height
B = area of the base
r = radius
h = height
$\pi \approx 3.14$

Circumference

Circle	$C = 2\pi r$ or $C = \pi d$

Volume

Prism or cylinder	$V = Bh$
Pyramid or cone	$V = \frac{1}{3}Bh$
Sphere	$V = \frac{4}{3}\pi r^3$

Area

Triangle	$A = \frac{1}{2}bh$
Rectangle or parallelogram	$A = bh$
Rhombus	$A = \frac{1}{2}d_1d_2$
Trapezoid	$A = \frac{1}{2}(b_1 + b_2)h$
Regular polygon	$A = \frac{1}{2}aP$
Circle	$A = \pi r^2$

Surface area

	Lateral	**Total**
Prism	$S = Ph$	$S = Ph + 2B$
Pyramid	$S = \frac{1}{2}Pl$	$S = \frac{1}{2}Pl + B$
Cylinder	$S = 2\pi rh$	$S = 2\pi rh + 2\pi r^2$
Cone	$S = \pi rl$	$S = \pi rl + \pi r^2$
Sphere		$S = 4\pi r^2$

Chapter 10

PRACTICE TEST 1

Directions: Read each question carefully. Write your answers on a separate sheet of paper. For multiple-choice questions, determine the best answer to the question from the four answer choices provided, and write the letter of the correct answer. For a griddable question, determine the best answer to the question, and write the numeric answer.

1. In which of the following steps does a mistake first appear in simplifying the expression $3(k + 4) - 7(2k - 1)$?

Step 1: $3k + 12 - 7(2k - 1)$

Step 2: $3k + 12 - 14k - 7$

Step 3: $3k - 14k + 12 - 7$

Step 4: $-11k + 5$

A. Step 1
B. Step 2
C. Step 3
D. Step 4

2. Identify the axis of symmetry and vertex of the quadratic function

$$y = x^2 - 4x + 6$$

A. $x = 2$ and (2, 2)
B. $x = 2$ and (2, 4)
C. $x = -2$ and (–2, 18)
D. $x = -2$ and (–2, 2)

3. What is the rate of change of the following table of values?

x	–5	0	5	10
y	5	9	13	17

A. $\frac{4}{5}$

B. $-\frac{4}{5}$

C. $\frac{5}{4}$

D. $-\frac{5}{4}$

4. Identify the range of the function graphed below.

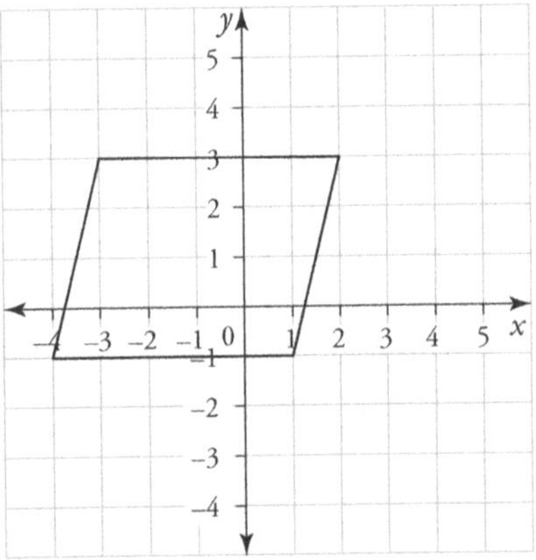

A. $-4 \le x \le 2$
B. $-4 \le y \le 2$
C. $-1 \le x \le 3$
D. $-1 \le y \le 3$

5. Find the 7th term of the sequence $a_n = 2^n - 1$.

Record your answer on your answer sheet.

6. Sally is helping sell cookies for her niece's t-ball league fundraiser. She has already sold 24 boxes of cookies and has started selling them in bundles of four. Which function can be used to find the total number of cookies she will sell after selling b more bundles.

   A. $h(b) = 24b + 4$
   B. $h(b) = 28b$
   C. $h(b) = 24b - 4b$
   D. $h(b) = 4b + 24$

7. Paul deposits money into a savings account for college and continues to add an equal amount every month thereafter. The table below shows the amount of money in his savings account for each corresponding month after he starts saving. ***Note:*** Earned interest is not included.

Number of Months	**Money (in Dollars)**
2	225
4	315
6	405
8	495

Based on this information, which statement is true?

   A. Paul initially deposited \$45 into his account.
   B. Paul initially deposited \$90 into his account.
   C. Paul initially deposited \$135 into his account.
   D. Paul initially deposited \$225 into his account.

8. Which table shows the same relationship as $y = 6x + 19$?

A.

x	–3	–1	2	3
y	–37	–25	–7	–1

B.

x	–1	2	3	6
y	13	31	37	55

C.

x	0	1	3	5
y	19	13	1	–11

D.

x	–4	–3	–1	0
y	–70	–51	–13	6

9. The first four numbers in a pattern are shown below.

$$-87, -75, -27, 165, \ldots$$

If the pattern continues, which expression can be used to find the *nth* term in the sequence?

A. $4^n - 91$

B. $12n - 87$

C. $4^n - 87$

D. $12n - 91$

10. Which of the following is a solution to $x^2 + 4x - 5 = 0$?

A. 1

B. –1

C. 5

D. –4

11. Factor the expression $4x^2 - 25$.

A. $(4x + 5)(x - 5)$

B. $(4x - 5)(x + 5)$

C. $(2x - 5)^2$

D. $(2x + 5)(2x - 5)$

12. The following tables represent the values for two linear functions in a system of equations. Based on the data, what is the solution of the system of equations?

x	*y*
–2	–2
0	2
2	6
4	10
5	12

x	*y*
–2	4
0	5
2	6
3	6.5
5	7.5

A. (–2, 4)

B. (0, 2)

C. (2, 6)

D. (4, 10)

13. Which of the following linear functions describes the linear parent function shifted down 4 units with a slope that is twice as steep?

A. $y = -2x + 4$

B. $y = 2x - 4$

C. $y = 2x + 4$

D. $y = -2x - 4$

14. Which of the following describes the slope of a line, m, that contains the points (a, b) and (a, c)?

A. $m = \frac{c}{b}$

B. $m = a$

C. zero slope

D. undefined slope

15. Which of the following sets of ordered pairs represents y as a function of x?

A. $\{(9, 5), (-2, 6), (-2, 5), (6, 5)\}$

B. $\{(4, -6), (3, 5), (-1, 0), (7, -2)\}$

C. $\{(5, 5), (-4, 5), (5, 4), (0, 5)\}$

D. $\{(1, -1), (1, -2), (1, 2), (1, -3)\}$

16. Which of the following is an equation of a line that's perpendicular to $y = 12$ and passes through the point $(-5, 24)$?

A. $y = 24$

B. $y = -5x + 24$

C. $x = -5$

D. $x = 24$

17. Which of the following graphs represents the inequality $8x + 4y \geq 12$?

A.

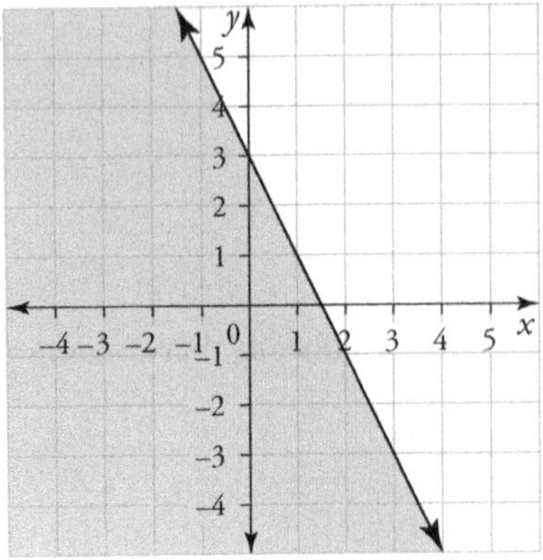

B.

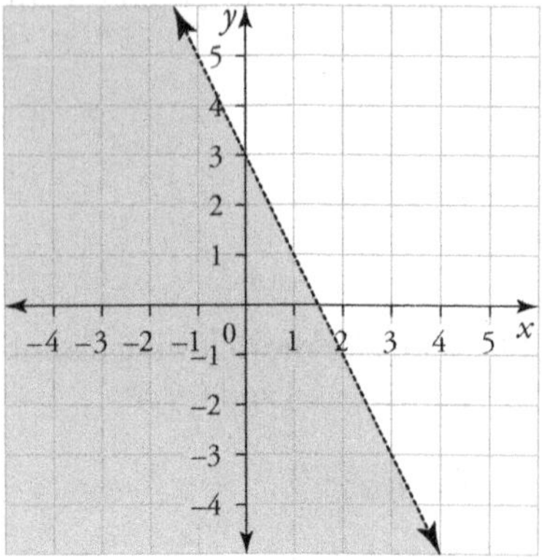

C.

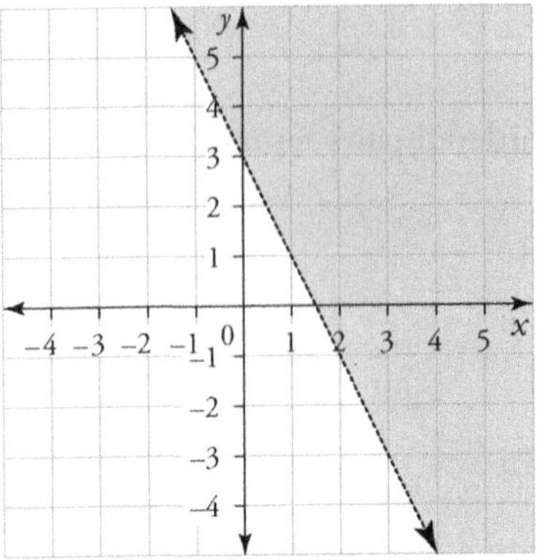

D.

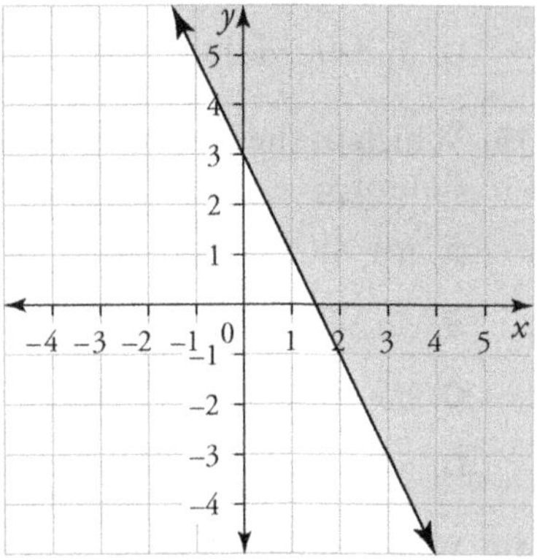

18. Identify the slope of the function $6x - 3y = 15$.

Record your answer on your answer sheet.

19. The graph of line h is represented by $y = 2x + 9$. If the slope of line h is multiplied by $\frac{-1}{2}$ to create line g, which statement about the graphs of the two lines is true?

A. Line g is steeper than line h.

B. Line h is steeper than line g.

C. Line g is parallel to line h.

D. Line g is perpendicular to line h.

20. Graph *d* represents the function $y = x^2 - 6$. If graph *f* is equal to $2y$, which of the following statements is an accurate description of the two graphs?

A. Graph *f* is translated up 2 units.

B. Graph *f* is translated to the right 2 units.

C. Graph *f* is wider than graph *d*.

D. Graph *f* is narrower than graph *d*.

21. The value of *y* varies directly with *x*. Which function represents the relationship between *x* and *y* if $y = 78$ when $x = -6$?

A. $x = -13y$

B. $x = \frac{y}{-13}$

C. $y = 13x$

D. $y = \frac{13}{x}$

22. Which expression represents the perimeter of this rectangle?

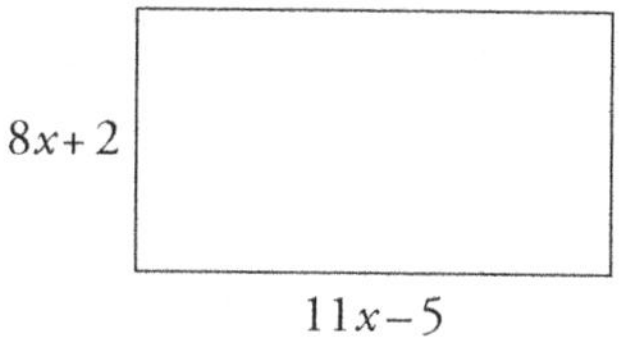

A. $19x - 3$

B. $38x - 6$

C. $88x - 10$

D. $38x + 6$

23. What is the equation in standard form of the line that passes through the point (8, 11) and has a slope of 0.5?

A. $x - 2y = 14$

B. $x + 2y = 88$

C. $x - 2y = 19$

D. $-x + 2y = 14$

24. Evaluate the function $g(h) = 3h^2 - 5$ for $h = -4$.

Record your answer on your answer sheet.

25. Which graph does not represent y as a function of x?

A.

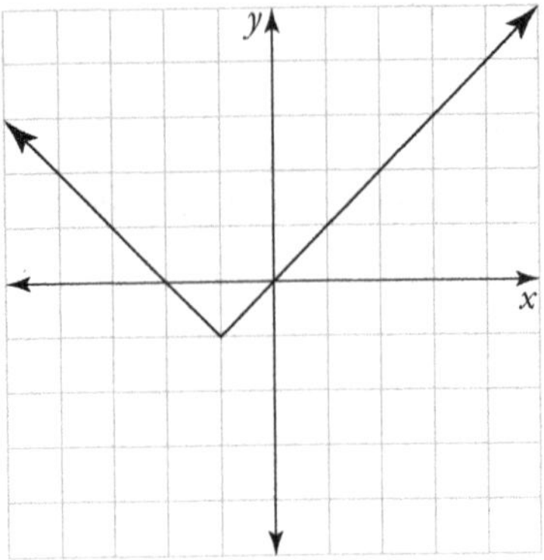

B.

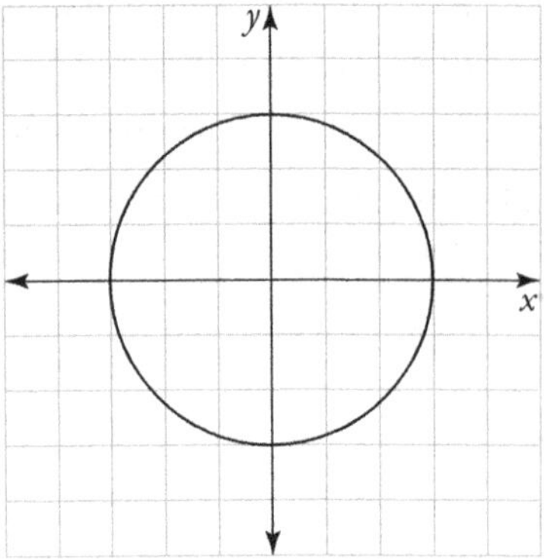

C.

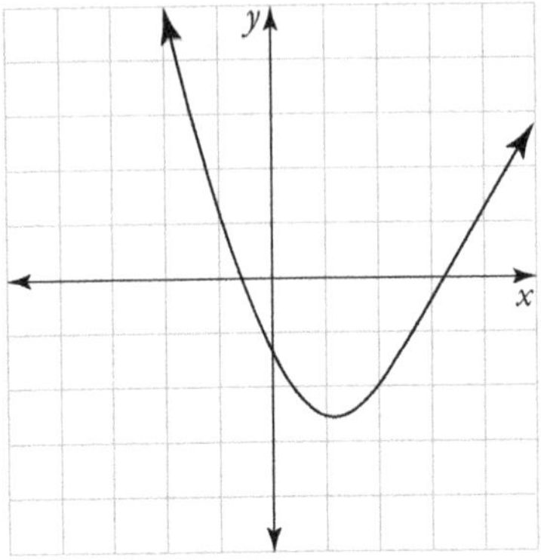

D.

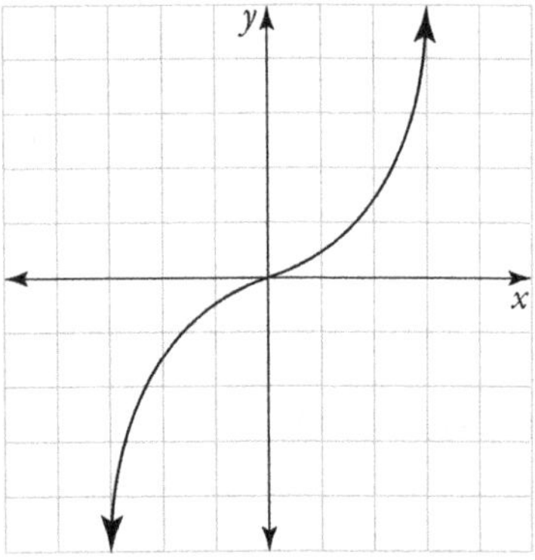

26. What is the domain of the function shown below?

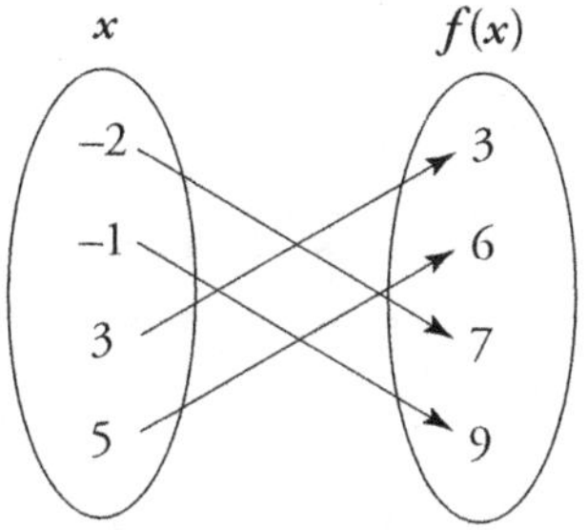

A. {3, 6, 7, 9}

B. {–2, –1, 3, 6, 5, 7, 9}

C. {3}

D. {–2, –1, 3, 5}

27. What is the slope of the linear function $12x - 7y = 28$?

A. $\frac{12}{7}$

B. $-\frac{12}{7}$

C. –4

D. 12

28. What type of correlation is shown in the graph below?

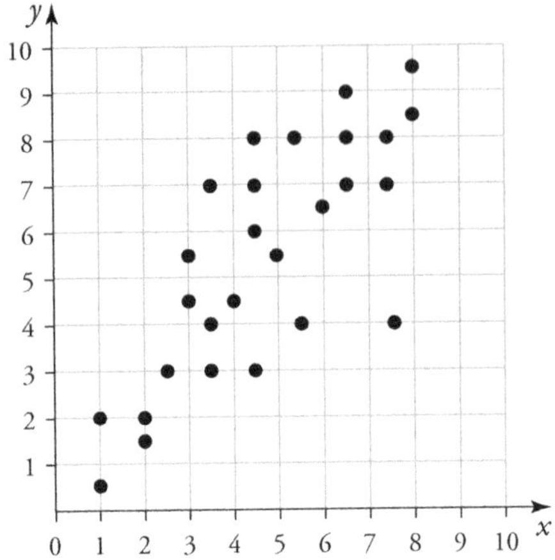

A. positive correlation

B. negative correlation

C. no correlation

D. inverse correlation

29. Joe's Coffee Shop has a frequent customer card that allows you to earn a free cup of coffee for every 5 cups that you buy. The function $y = \frac{1}{5}x$ represents *y*, the number of free cups of coffee you get, in terms of *x*, the number of cups of coffee you bought. You went to Joe's Coffee Shop 20 times in one month. What is a reasonable range for this function?

A. $\{0, 1, 2, 3, 4\}$

B. $\{4, 5, 20\}$

C. $0 < y < 4$

D. $0 < y < 20$

30. Find the 16th term of the sequence $a_n = 2n + 3$.

A. 5

B. 6.5

C. 38

D. 35

31. Which of the following equations is represented by the graph below?

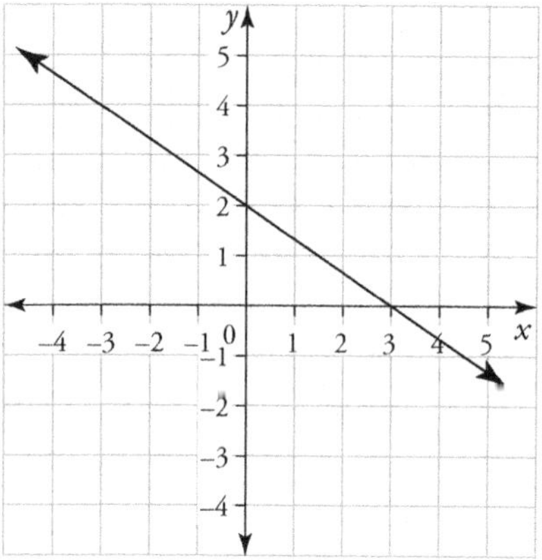

I. $2x + 3y = 6$

II. $y = -\frac{2}{3}x + 2$

III. $y - 4 = -\frac{2}{3}(x + 3)$

A. I and II

B. II and III

C. I and III

D. All of the above

32. Which expression is equivalent to $6x^2 + 5x - 4$?

A. $(6x + 2)(x - 2)$

B. $(3x - 1)(2x + 4)$

C. $(6x - 2)(x + 2)$

D. $(3x + 4)(2x - 1)$

33. Which of the following is a solution to the system of equations modeled by the graph below?

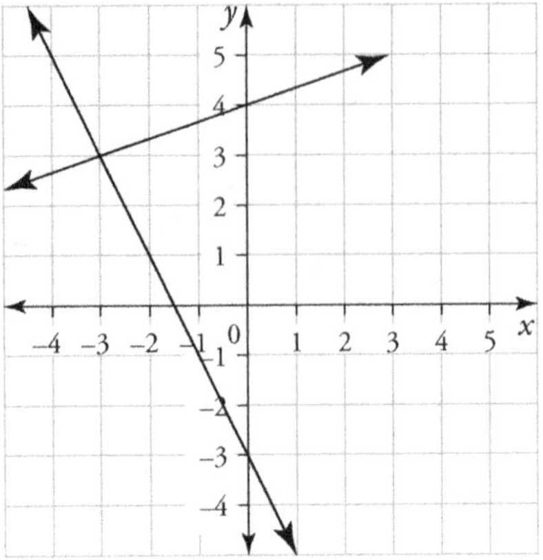

A. (3, –3)
B. (0, 4)
C. (–3, 3)
D. (0, –3)

34. The cross-country team has 50 tickets to sell for their annual fundraiser dinner. Each adult ticket costs \$5, and each student ticket costs \$2. Which system of equations can be used to find the number of adult tickets, *y*, and student tickets, *x*, needed to sell \$160 worth of tickets?

A. $x + y = 160$
$2x + 5y = 50$
B. $x + y = 50$
$5x + 2y = 160$
C. $x + y = 160$
$5x + 2y = 50$
D. $x + y = 50$
$2x + 5y = 160$

35. Which of the following is a solution to $y > -\frac{1}{4}x + 10$?

A. (–8, 12)
B. (16, 5)
C. (13, 9)
D. (–24, 11)

36. The first side of a triangle has a length of $(x - 4)$, the second side has a length of $2x$, and the third side has a length of 10. Find the value of x if the perimeter of the triangle is 36 inches.

Record your answer on your answer sheet.

37. Which of the following describes the range of the graph below?

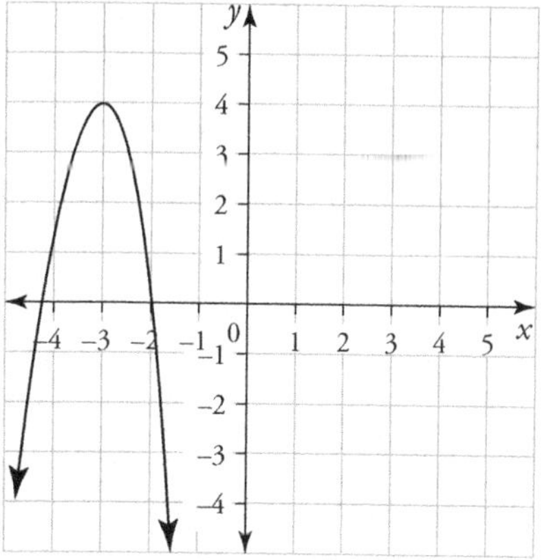

A. $-\infty < y < \infty$

B. $-5 \leq x \leq -1$

C. $y \leq 4$

D. $x \geq -4$

38. Solve $f(x) = -9x - 3$ when $f(x) = -12$.

A. 1

B. 105

C. 0

D. –1

39. Which of the following expressions is equivalent to $2x^2 - 13x - 24$?

A. $(2x - 8)(x + 3)$

B. $(2x - 3)(x + 8)$

C. $(2x + 3)(x - 8)$

D. $(2x - 3)(x - 8)$

40. If $f(x)$ represents the graph of the quadratic parent function, describe the effects of graphing $f(x - h)$.

A. The graph is translated h units to the left.
B. The graph is translated h units to the right.
C. The graph is translated h units up.
D. The graph is translated h units down.

41. Evaluate $f(x) = 8x^2 + 2x - 19$ for $x = -3$.

Record your answer on your answer sheet.

42. Which equation is equivalent to $6x + 2y = 18$?

A. $y = 3x + 9$
B. $y = -3x + 9$
C. $y = 3x - 9$
D. $y = -3x - 9$

43. A table of values for an exponential function is shown below.

x	y
0	400
1	392
2	384.16
3	376.48

Which situation could describe this function?

A. The price of a smart tablet decreases \$8 each year.
B. The price of a smart tablet decreases by 2% each year.
C. The price of a smart tablet increases by \$8 each year.
D. The price of a smart tablet increases by 2% each year.

44. The area of a rectangle is represented by the expression $9x^2 - 16$. Which of the following expressions could represent the length and width of this rectangle?

A. $(3x + 4)(3x + 4)$
B. $(3x + 4)(3x - 4)$
C. $(9x + 8)(x - 2)$
D. $(9x - 2)(x + 8)$

45. Graph d represents the function $y = x^2 - 6$. If graph f is equal to $\frac{1}{2}y$, which of the following statements is an accurate description of the two graphs?

A. The graph f is translated up 2 units.

B. The graph f is translated to the right 2 units.

C. The graph f is wider than graph d.

D. The graph f is narrower than graph d.

46. Which expression is equivalent to $(6x^3y^5)^{-2}$?

A. $-12xy^3$

B. $\frac{-12}{xy^3}$

C. $\frac{1}{36x^6y^{10}}$

D. $\frac{36}{x^6y^{10}}$

47. Which equation shows the same relationship as the table of values below?

x	0	1	2	3	4
y	2	6	18	54	162

A. $y = 3 \cdot 2^x$

B. $y = x^2 + 2$

C. $y = 2 \cdot 3^x$

D. $y = x + 2$

48. The area of a rectangle is represented by the expression $9x^2 + 24x + 16$. Which of the following expressions could represent the length and width of this rectangle?

A. $(3x + 4)(3x + 4)$

B. $(9x - 2)(x + 8)$

C. $(3x + 4)(3x - 4)$

D. $(9x + 8)(x - 2)$

49. When released into the atmosphere, the radioactive isotope seaborgium-226 has a half-life of 30 seconds. If 4 mg is released into the atmosphere, which of the following functions describes the amount of seaborgium-226 remaining after m number of minutes?

A. $f(m) = 226 - 0.5m$

B. $f(m) = 4(0.5)^m$

C. $f(m) = 4 - 0.05m$

D. $f(m) = 4(0.25)^m$

50. What is the vertex of the graph of the quadratic function $f(x) = (x - 3)^2 + 5$?

A. (–3, 5)

B. (3, 0)

C. (0, 5)

D. (3, 5)

51. The function $p = 82{,}000(1.2)^t$ models the population of a town over time. Identify what the value 1.2 represents in this equation.

A. population size of the town after t number of years

B. initial population size of the town

C. growth rate at which the population is expanding

D. number of years

52. The height of a coin thrown off of a 50-foot building at a speed of 32 feet per second is described by the function $h(t) = -16t^2 + 32t + 50$, where t represents the number of seconds after the coin is thrown. Approximately how much time will it take for the coin to hit the ground?

A. between 0 and 1 second

B. between 1 and 2 seconds

C. between 2 and 3 seconds

D. between 3 and 4 seconds

53. What is the zero of the linear function graphed below?

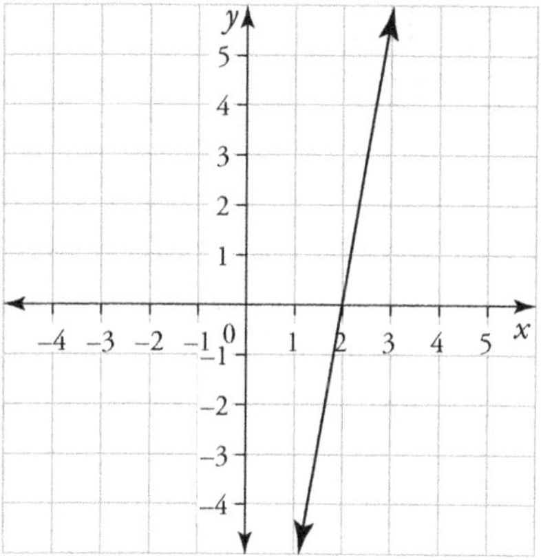

Record your answer on your answer sheet.

54. The following graph represents the relationship between the number of rounds in an elimination tournament and the number of players remaining. Identify a reasonable range of this function.

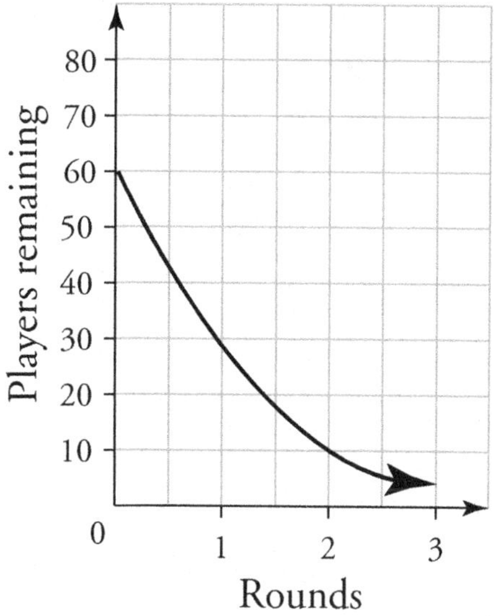

A. $1 \leq y \leq 64$

B. $1 \leq y \leq 6$

C. $1 \leq x \leq 64$

D. $1 \leq x \leq 6$

Answers

1. B
2. A
3. A
4. D
5. 127
6. D
7. C
8. B
9. A
10. A
11. D
12. C
13. B
14. D
15. B
16. C
17. D
18. 2
19. B
20. D
21. B
22. B
23. D
24. 43
25. B
26. D
27. A
28. A
29. A
30. D
31. D
32. D
33. C
34. D
35. C
36. 10
37. C
38. A
39. C
40. B
41. 47
42. B
43. B
44. B
45. C
46. C
47. C
48. A
49. D
50. D
51. C
52. D
53. 2
54. A

Chapter 11

PRACTICE TEST 2

Directions: Read each question carefully. Write your answers on a separate sheet of paper. For multiple-choice questions, determine the best answer to the question from the four answer choices provided, and write the letter of the correct answer. For a griddable question, determine the best answer to the question, and write the numeric answer.

1. Talon is looking to buy some video games online. A website charges \$42.95 per video game, plus \$4.50 for shipping and handling. Talon has a total of \$135 to spend on video games. Which inequality represents the total number of video games, *g*, Talon can purchase?

A. $42.95g + 4.5 > 135$

B. $42.95g + 4.5 \leq 135$

C. $42.95g - 4.5 < 135$

D. $38.45g \leq 135$

2. Given the two functions below, how does the graph of *f* compare to the graph of *g*?

$$f(x) = \frac{1}{2}x - 5$$
$$g(x) = 2x + 5$$

A. The graph of *f* is steeper than the graph of *g*.

B. The two graphs are perpendicular.

C. The two graphs share a *y*- intercept.

D. The graph of *f* is less steep than the graph of *g*.

3. Which of the following graphs represents the solution to the following system of equations?

$$y = -3$$
$$y = 2x - 1$$

A.

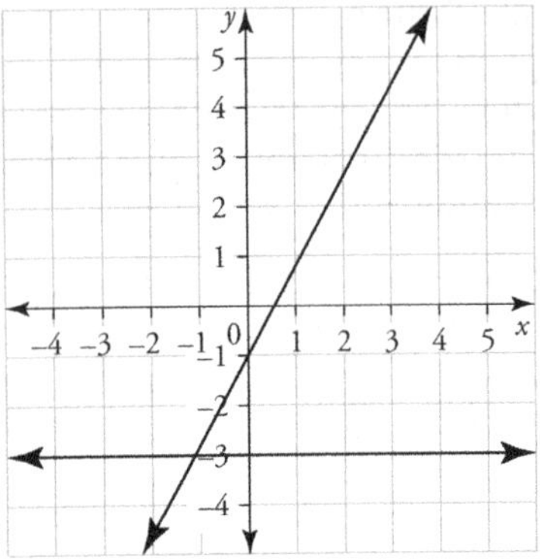

B.

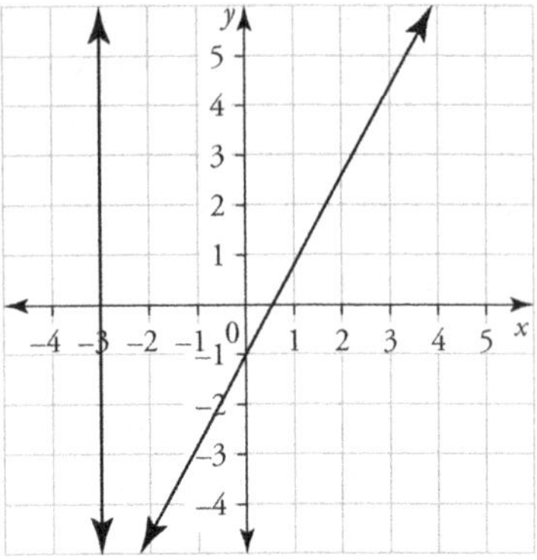

C.

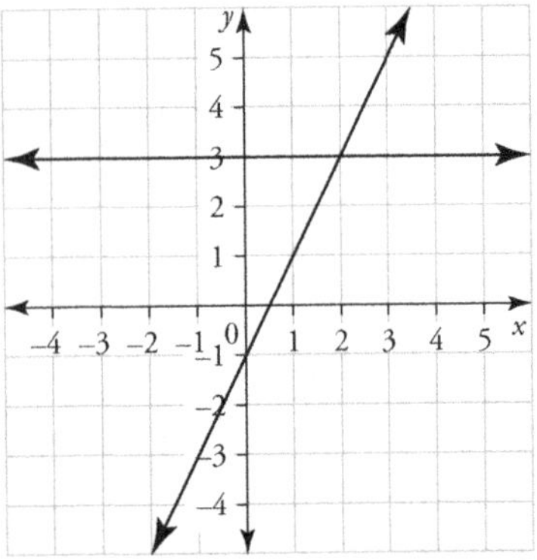

D.

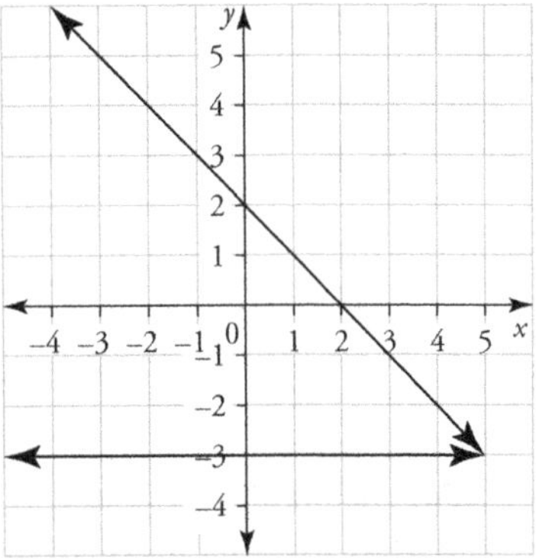

4. A car rental company charges you an initial fee of $39.99 to rent a car and allows you to travel 100 miles. The company uses the function $C(m) = 0.10(m - 100) + 39.99$ to calculate the total cost. Which of the following statements is true?

A. Every mile costs an additional $0.10.

B. The miles do not affect the total cost.

C. The most you will be charged is $100.

D. Every mile over 100 costs an additional $0.10.

5. Evaluate the function $f(x) = 2(x + 1)^2$ for the domain $\{-1, 0, 2, 3\}$.

A. $\{0, 2, 18, 32\}$

B. $\{-1, 1, 5, 7\}$

C. $\{1, 1, 25, 49\}$

D. $\{-0.5, -1, 0, 2.2\}$

6. Betty and Martha are both driving 530 miles from El Paso to Austin. At a certain time, Betty has already driven 45 miles and continues to travel at a constant rate of 60 mph. Martha has already driven 135 miles and continues to travel at a constant rate of 50 mph. Their distance from El Paso can be represented by the functions $f(x) = 60x + 45$ and $g(x) = 50x + 135$ respectively, where x represents the number of hours driven. How far apart, in miles, will they be from each other 4 hours later?

Record your answer on your answer sheet.

7. Which of the following is an equation of a line that's perpendicular to $-x + 2y = 6$ and passes through the point (3, 8)?

A. $2x + y = 14$

B. $-2x - y = 8$

C. $2x + y = 3$

D. $-2x + y = 6$

8. The first four numbers in a pattern are shown below.

2, 5, 10, 17, ...

If the pattern continues, which expression can be used to find the nth term in the sequence?

A. $\frac{n^2}{1+n}$

B. $\frac{n^3}{n-1}$

C. $\frac{n^2}{1-n}$

D. $n^2 + 1$

9. Choose the correct quadratic function that is described by the following statements.

1. The graph has a maximum.
2. The axis of symmetry is $x = 2$.
3. The equation has no solution.

A. $0 = x^2 + 2x - 6$
B. $0 = -2x^2 + 8x - 12$
C. $0 = 2x^2 + 8x - 6$
D. $0 = -3x^2 + 12x - 4$

10. Which table shows a rate of change of -3?

A.

x	–3	–2	–1	0
y	–8	–5	–2	1

B.

x	0	3	6	9
y	–3	0	3	6

C.

x	–3	0	1	4
y	11	2	–1	–10

D.

x	0	3	6	9
y	3	0	–3	–6

11. Which of the following is an equation of a line that's parallel to $y = -11x + 3$ and passes through the point $(1, -4)$?

A. $-11x + y = 7$
B. $11x + y = 7$
C. $11x + 7 = -4$
D. $-11x + y = 1$

12. In the graph of the quadratic function shown, one of the zeros is between which two *x* values?

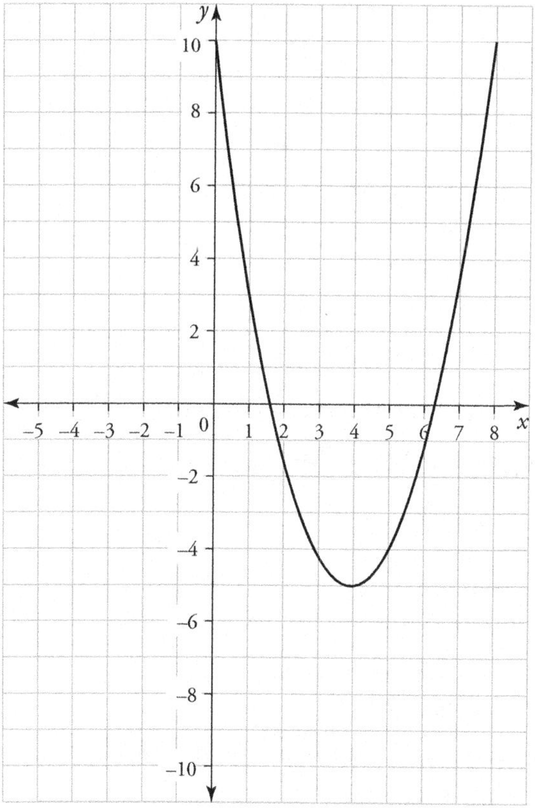

A. 3 and 4

B. 0 and 1

C. 1 and 2

D. 5 and 6

13. What is the zero of the linear function graphed below?

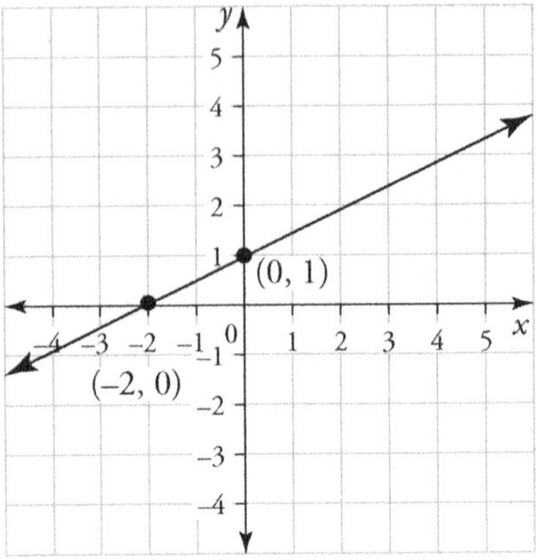

Record your answer on your answer sheet.

14. Which expression is equivalent to $x^2 - 26x + 169$?

A. $(x - 13)^2$
B. $(x + 13)^2$
C. $(x + 13)(x - 13)$
D. $(x - 26)(x + 169)$

15. If $f(x)$ represents the graph of the quadratic parent function, describe the effects of graphing $f(x + h) - k$.

A. The graph is translated h units to the left and k units up.
B. The graph is translated h units to the left and k units down.
C. The graph is translated h units to the right and k units up.
D. The graph is translated h units to the right and k units down.

16. In which step does a mistake first appear in simplifying the expression $(z + 3)(z^2 + 4z - 9)$?

Step 1: $z(z^2 + 4z - 9) + 3(z^2 + 4z - 9)$
Step 2: $(z^3 + 4z^2 - 9z) + (3z^2 + 12z - 27)$
Step 3: $z^3 + 4z^2 + 3z^2 - 9z + 12z - 27$
Step 4: $5z^3 + 7z^4 + 3z^2 - 27$

A. Step 1
B. Step 2
C. Step 3
D. Step 4

17. Which of the following functions has a slope of 2?

A. $14x + 7y = 8$
B. $-14x - 7y = 13$
C. $-14x + 7y = 11$
D. $7x - 14y = 5$

18. If (15, 6) and (x, –4) are points on the graph of a direct variation, find the missing value.

A. 2
B. 5
C. –10
D. –5

19. Which graph does NOT represent y as a function of x?

A.
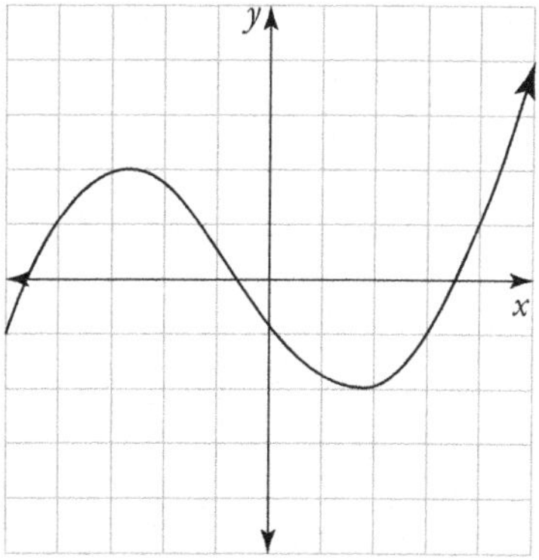

B.
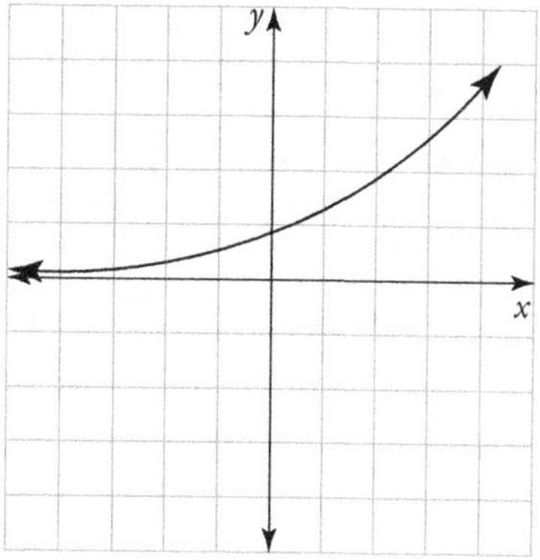

C.
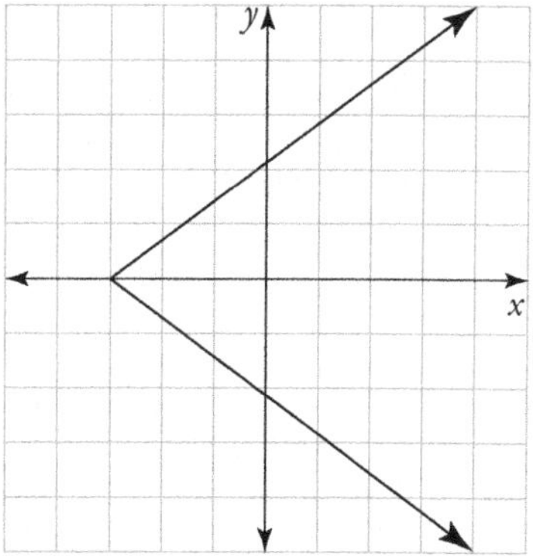

D.
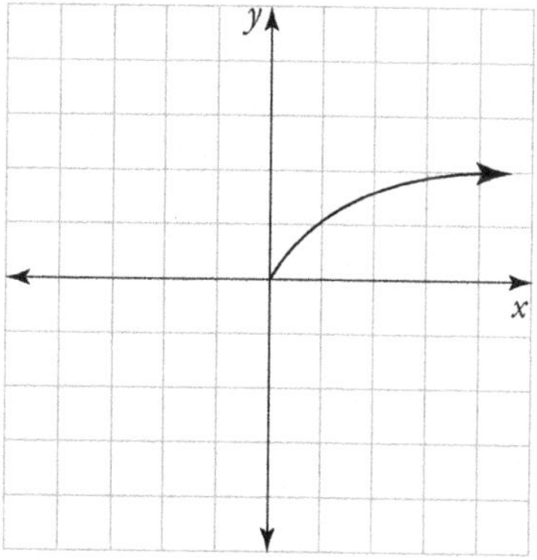

20. Identify the domain of the relation graphed below.

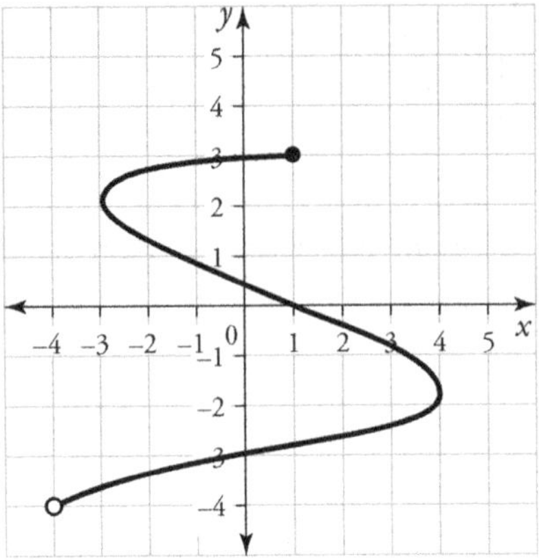

A. $-4 < x \leq 1$

B. $-4 < x \leq 4$

C. $-4 < y \leq 3$

D. $-4 < y \leq 4$

21. What is the range of the function shown below?

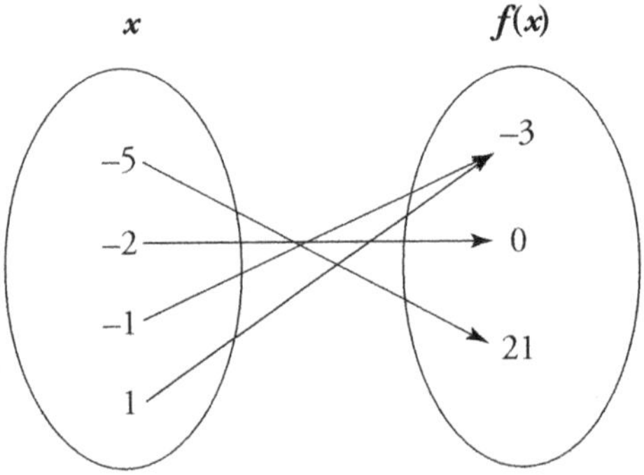

A. $\{-5, -2, -1, 1\}$

B. $\{-3, 0, 21\}$

C. $\{-5, -3, -2, -1, 0, 2, 21\}$

D. $\{-3\}$

22. Simplify $(3x - 4)(3x + 4)$.

A. $9x^2 - 16$

B. $9x^2 + 16$

C. $9x^2 + 8x - 16$

D. $9x^2 - 8x - 16$

23. A teacher is planning a pizza party for the end of the school year. He estimates that he will need $\frac{1}{4}$ of a pizza for every student, and he always orders 5 extra pizzas to share with the teachers next door. Which function can be used to find the total number of pizzas he will need to order for n students?

A. $g(n) = \frac{1}{4}n + 5$

B. $g(n) = 5n + \frac{1}{4}$

C. $g(n) = 5.25n$

D. $g(n) = 5 - \frac{1}{4}n$

24. The formula for converting a temperature in degrees Fahrenheit to degrees Celsius is $°C = (°F - 32)\frac{5}{9}$. Which of the following equations represents the formula for converting degrees Celsius to degrees Fahrenheit?

A. $°F = \frac{5}{9}°C + 32$

B. $°F = \frac{5}{9}°C - 32$

C. $°F = \frac{9}{5}°C + 32$

D. $°F = \frac{9}{5}°C - 32$

25. Which of the following relations represent y as a function of x?

A. {(2, 3), (3, 4), (4, 5), (5, 6)}

B.

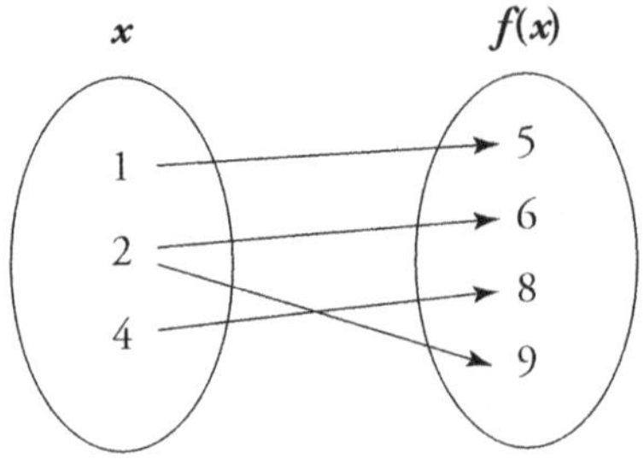

C.

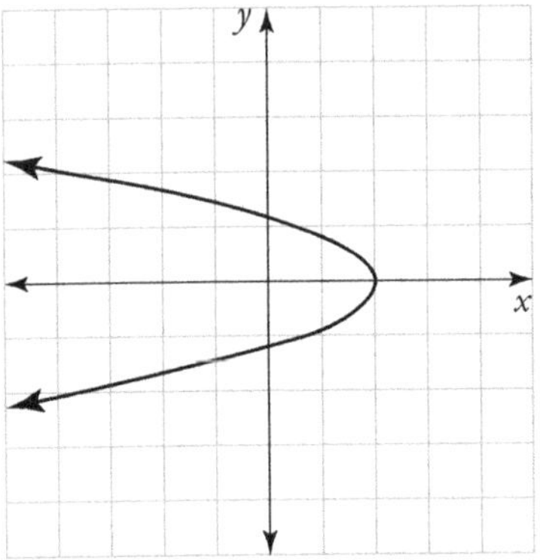

D.

x	y
–2	–1
–3	0
–2	1
–1	2

26. Oscar owns a professional landscaping business. He can complete between one and five landscaping jobs per month. What is a reasonable range for the number of landscaping jobs Oscar can complete in the next 2 months?

A. {1, 2, 3, 4, 5}

B. $1 \leq x \leq 5$

C. {1, 2, 3, 4, 5, 6, 7, 8, 9, 10}

D. $1 \leq y \leq 10$

27. The first four terms of a sequence are listed below.

$$-0.25, 0.25, 0.75, 1.25$$

Which of the following expressions describes the nth term of the sequence?

A. $n - 0.75$

B. $0.25n - 0.5$

C. $-n + 0.75$

D. $0.5n - 0.75$

28. The following graph represents the relationship between the number of hours a student studies and his or her respective grade on a mathematics test. Identify a reasonable range of this function.

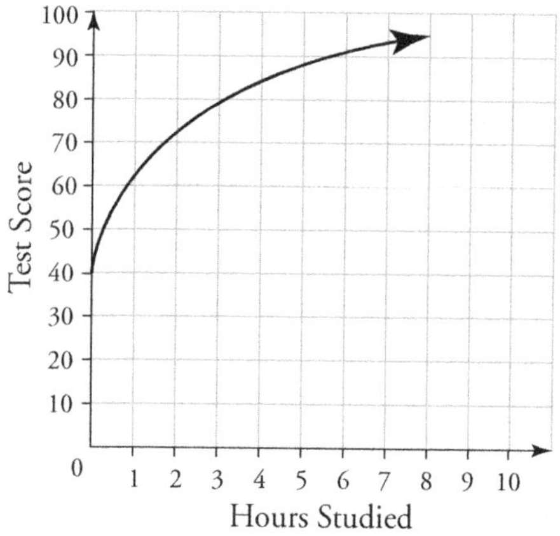

A. $0 \le x \le 10$

B. $40 \le x \le 100$

C. $0 \le y \le 10$

D. $40 \le y \le 100$

29. What is the equation of the line that passes through (2, 13) and has a slope of –4?

A. $y = -4x - 21$

B. $y = -4x + 21$

C. $y = 4x + 21$

D. $y = 4x - 21$

30. Which of the following equations are represented by the graph below?

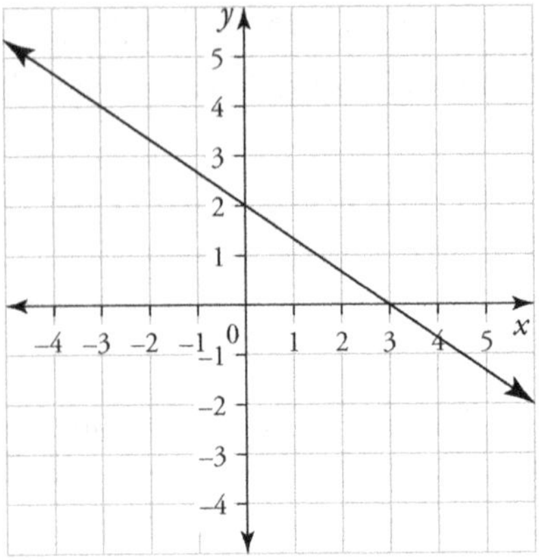

I. $2x - 3y = 6$

II. $y = -\frac{2}{3}x + 2$

III. $y - 4 = -\frac{2}{3}(x + 3)$

A. I and II

B. II and III

C. I and III

D. All of the above

31. Which of the following equations is represented by the graph below?

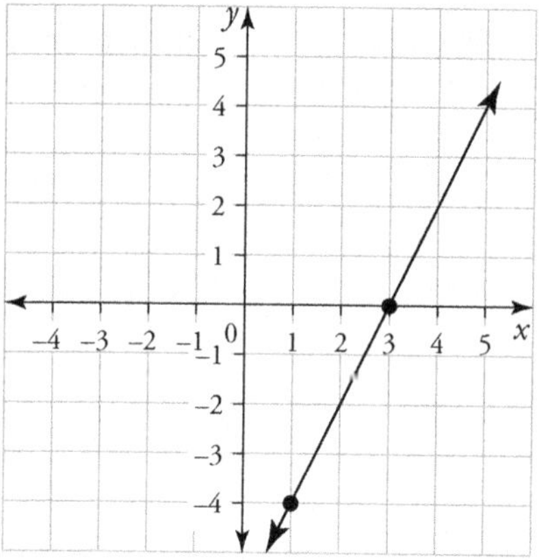

A. $y - 4 = 2(x + 1)$
B. $y - 3 = 2(x + 1)$
C. $y - 4 = 2(x + 3)$
D. $y + 4 = 2(x - 1)$

32. Mr. Alexander has graded 39 algebra tests by the end of his conference period. He knows he can continue to grade, on average, around 26 tests every additional hour. Which graph models the total number of tests Mr. Alexander has graded over a certain amount of time after his conference period?

A.

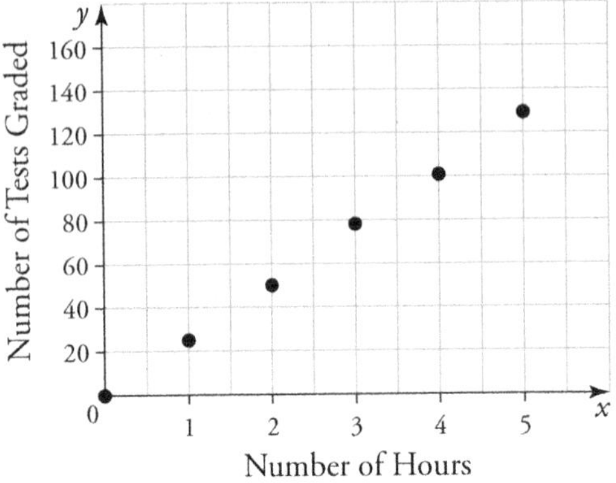

B.

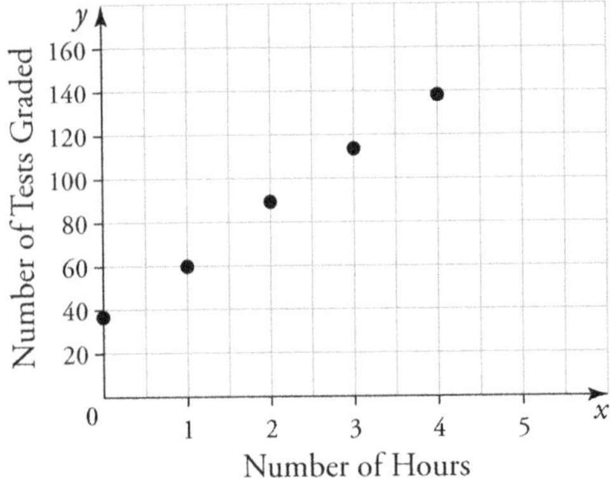

C.

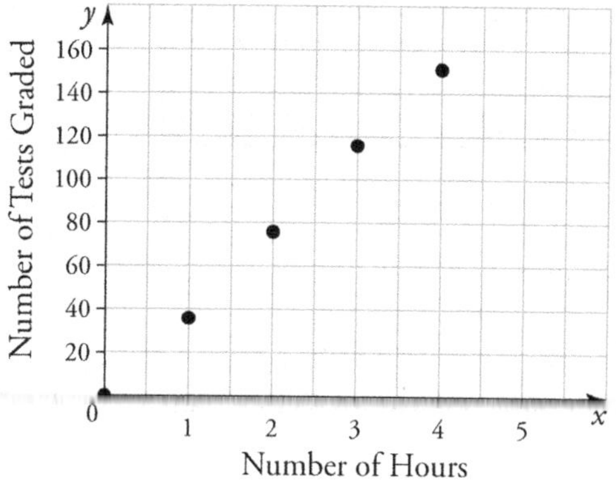
y
160
140
120
100
80
60
40
20
0
1
2
3
4
5
x
Number of Tests Graded
Number of Hours

D.

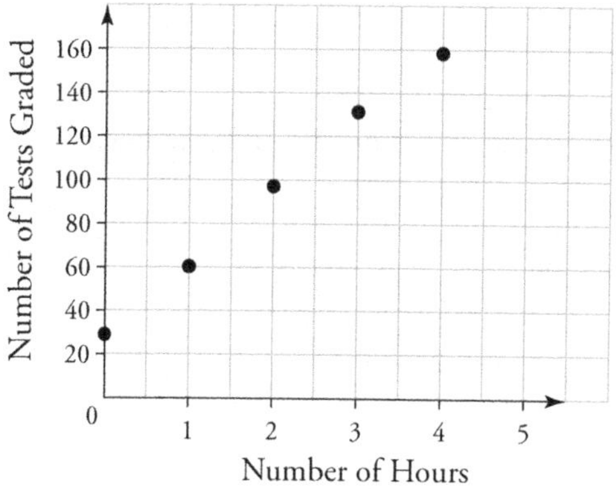
160
140
120
100
80
60
40
20
0
1
2
3
4
5
Number of Tests Graded
Number of Hours

33. Which equation shows the same relationship as the table below?

x	–5	–2	0	1
$f(x)$	27	3	–13	–21

A. $f(x) = -24x - 13$
B. $f(x) = -8x - 13$
C. $f(x) = 8x - 13$
D. $f(x) = 24x - 5$

34. If $y = 13x - 21$, what is the value of x when $y = -34$?

A. –463
B. 421
C. –1
D. –4

35. Which expression represents the perimeter of this triangle?

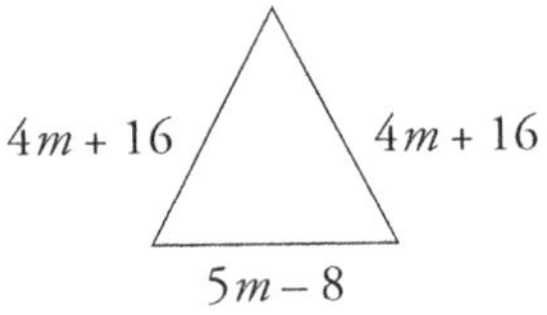

A. $13m + 24$
B. $13m + 8$
C. $20m + 8$
D. $20m + 24$

36. What is the value of x in the solution to the system of equations below?

$$3x - 2y = 18$$
$$2x + y = 5$$

A. –3
B. 4
C. 3
D. 6

37. Which of the following describes the domain of the graph below?

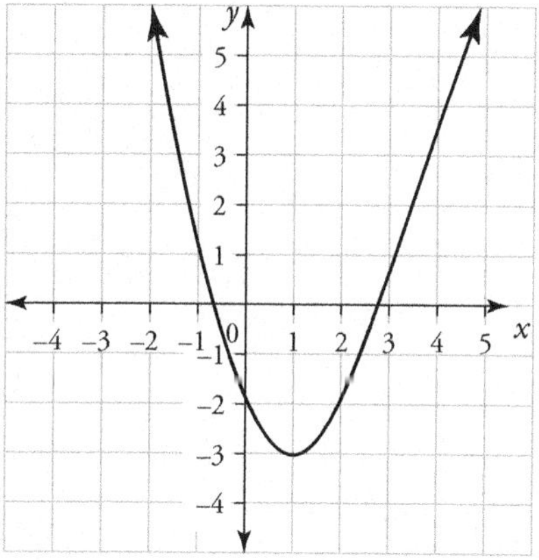

A. $y \geq -3$

B. $-2 \leq x \leq 4$

C. $-\infty < y < \infty$

D. $-\infty < x < \infty$

38. Which inequality can be represented by the graph below?

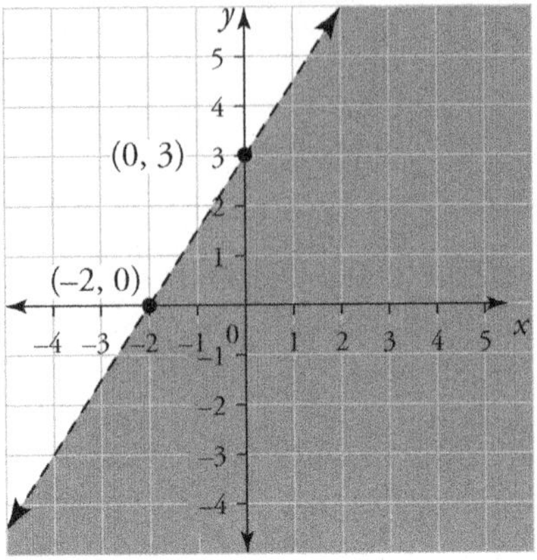

A. $y < \frac{3}{2}x - 2$

B. $y < \frac{2}{3}x + 3$

C. $y < \frac{3}{2}x + 3$

D. $y < -\frac{2}{3}x + 3$

39. James invites a group of his friends along for dinner at a restaurant. Some of his friends bring their kids; all together, there are 8 people in their group. Items on the kid's menu cost \$6, and items on the adult menu are \$9 each. The total bill, before tax and tip, is \$63. Which of the following systems of equations represents the number of adults, a, and the number of kids, k, that were in James' dinner group?

A. $a + k = 63$
$6k + 9a = 8$

B. $a + k = 8$
$9k + 6a = 63$

C. $a + k = 63$
$9k + 6a = 8$

D. $a + k = 8$
$6k + 9a = 63$

40. The student council has to bake 3,000 cookies for their annual bake sale. They have already baked 600 cookies and can bake 300 cookies per hour. The total number of cookies that they bake after x hours can be represented by the function $y = 300x + 600$. They decide to have the PTA help bake cookies, which will double the number of cookies they can bake per hour. What statement describes the changes that were made to the graph of the original function?

A. The line will be translated up.

B. The line will be translated down.

C. The line will become less steep.

D. The line will become steeper.

41. At a certain high school, there are a total of 1,253 students. If there are 57 more girls than boys, what is the total number of boys at the school?

Record your answer on your answer sheet.

42. The value of y varies directly with x. Which function represents the relationship between x and y if $y = -28$ when $x = -4$?

A. $x = \frac{y}{7}$

B. $x = 7y$

C. $y = \frac{7}{x}$

D. $y = 7x$

43. Identify the table of values that models the equation $f(x) = 240(0.5)^x$.

A.

x	0	2	4	6
y	240	239	238	237

B.

x	0	1	4	9
y	0	240	480	720

C.

x	0	3	5	6
y	240	120	60	30

D.

x	1	2	4	6
y	120	60	15	7.5

44. Which of the following is the parent function of the relation modeled by this graph?

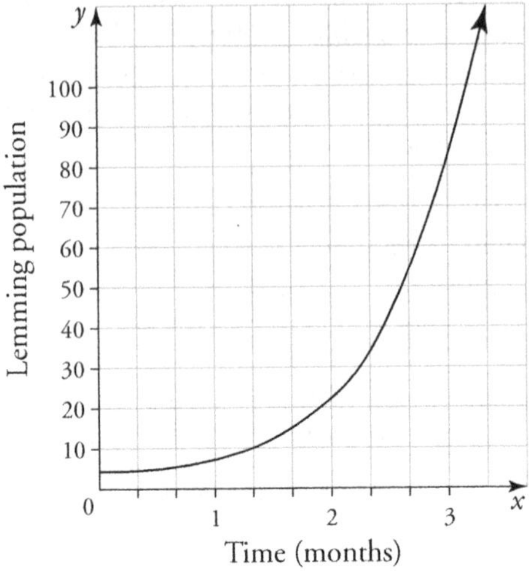

A. $f(x) = x$
B. $f(x) = x^2$
C. $f(x) = a \cdot b^x$
D. $f(x) = |x|$

45. Which expression is equivalent to $\dfrac{8a^3b^{-2}c^4}{2ab^{-5}c^6}$?

A. $\dfrac{6a^4c^{10}}{b^{-7}}$

B. $\dfrac{4a^4c^{10}}{b^{-7}}$

C. $\dfrac{6a^2b^3}{c^2}$

D. $\dfrac{4a^2b^3}{c^2}$

46. For the function $f(x) = 19x - 42$, find x when $f(x) = 129$.

Record your answer on your answer sheet.

47. Which of the following equations models the relationship between the size of a bacteria culture, b, that starts with 1,200 live bacteria and triples in size every hour, h?

A. $1{,}200 = 3h + b$

B. $b = 3h^2 + 1{,}200$

C. $b = 3(1{,}200)^h$

D. $b = 1{,}200(3)^h$

48. What type of correlation would you expect to see when comparing the speed of a runner and the time it takes her to reach the finish line?

A. positive correlation

B. negative correlation

C. constant correlation

D. no correlation

49. Approximate the value of the function $y = 5{,}200(0.75)^x$ for $x = 9$.

A. 69,255

B. 390

C. 35,100

D. 2.09×10^{22}

50. Which of the following expressions represents the area of a rectangle with a length of $(x - 3)$ and a width of $(x^2 + 4x - 5)$?

A. $x^3 + x^2 - 17x + 15$

B. $3x^2 - 12x + 15$

C. $x^2 + 4x - 8$

D. $x^3 + 4x^2 - 5x + 15$

51. How many solutions are there to the following system of equations?

$$y = \frac{2}{3}x - 2$$

$$8x - 12y = 24$$

A. one solution

B. two solutions

C. no solutions

D. infinitely many solutions

52. What is the domain of the function represented by this graph?

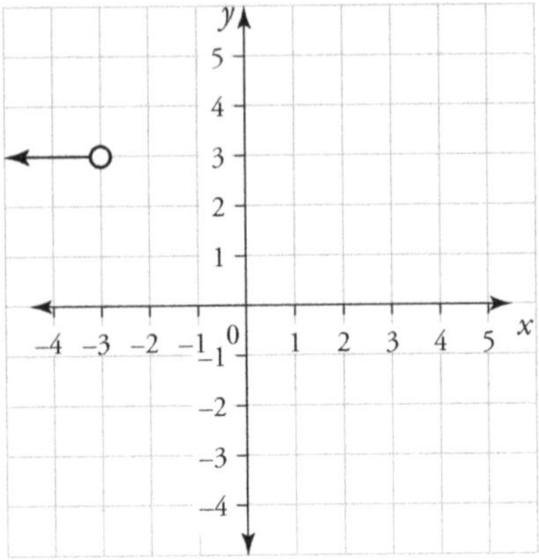

A. $x \leq -3$

B. $x < -3$

C. $y < -3$

D. $y \leq -3$

53. Which of the following describes the slope of a line, m, that contains the points (a, c) and (b, c)?

A. $m = \frac{c}{b}$

B. $m = a$

C. zero slope

D. undefined slope

54. Solve $g(x) = -3x^2 + 5x + 23$ for $x = -7$.

Record your answer on your answer sheet.

Answers

1. B
2. D
3. A
4. D
5. A
6. 50
7. A
8. D
9. B
10. C
11. B
12. C
13. –2
14. A
15. B
16. D
17. C
18. C
19. C
20. B
21. B
22. A
23. A
24. C
25. A
26. C
27. D
28. D
29. B
30. B
31. D
32. B
33. B
34. C
35. A
36. B
37. D
38. C
39. D
40. D
41. 598
42. D
43. D
44. C
45. D
46. 9
47. D
48. B
49. B
50. A
51. D
52. B
53. C
54. –159

Made in the USA
Monee, IL
04 November 2026

67384049R00154